Alexandre Tomoyose

Stereoscopy for Augmented Reality and Games

Alexandre Tomoyose

Stereoscopy for Augmented Reality and Games

Analysing, Comparing and Classifying Stereoscopic Techniques

ScienciaScripts

Imprint

Cover image: www.ingimage.com

This book is a translation from the original published under ISBN 978-3-330-76044-8.

Publisher:
Sciencia Scripts
is a trademark of
Dodo Books Indian Ocean Ltd. and OmniScriptum S.R.L publishing group

120 High Road, East Finchley, London, N2 9ED, United Kingdom
Str. Armeneasca 28/1, office 1, Chisinau MD-2012, Republic of Moldova, Europe
Printed at: see last page
ISBN: 978-620-8-33864-0

CONTENTS

CHAPTER 1

INTRODUCTION

Stereoscopy is a very broad area of knowledge, present in texts on medicine (ZYGMUNT et al., 2005), art (ADAMS, 2003) and technology (HOLLIMAN, 2002). In general, the term stereoscopy is only touched on in the texts that refer to it, without going into greater detail about its development. Works dedicated to in-depth research in this area are scarce (ADAMS, 2003).

Projects that explore the use of interactive virtual environments or augmented environments with the addition of real elements do not exploit the full potential for visualisation and depth perception that stereoscopy provides. Based on the concept of telepresence, which is the experience of being immersed in an environment through a communication medium (STEUER 1992), further exploration of stereoscopy could provide the necessary support for these projects to go one step further in the user's experience of telepresence, adding depth perception to the visualisation of interactive virtual and augmented environments.

As an example of this proposal, some of the projects developed at the Interactive Technologies Laboratory (INTERLAB) in the area of virtual and augmented reality can take advantage of the added value of stereoscopy. One of them is EnJine, which consists of a didactic game engine implemented on the Java and Java 3D platform (JAVA3D, 2010) that stands out for its organisation, which follows standards recommended by software engineering, and ease of learning and use (TORI et al., 2006). One of the tasks of the students on the computer graphics course is to use EnJine to develop a game and thus grasp basic design, engineering and computer graphics concepts.

However, the main project developed at INTERLAB that has potential for stereoscopy is Video-Avatar, part of FAPESP's TIDIA- AE project, which works on integrating avatars of real people, generated through videos, with a three-dimensional virtual environment generated by computer graphics. For this integration to be as realistic as possible, it is necessary to apply stereoscopic solutions that enable participants to visualise the environment and the other avatars in three dimensions. A first step in this direction was the Augmented Virtuality Tele-Conferencing (AVTC) developed at INTERLAB (SISCOUTTO, 2003). The AVTC system generates a virtual tele-conferencing environment with the addition of real video avatars and allows limited stereoscopic visualisation of video avatar images in a controlled environment.

Considering the range of resources that stereoscopy has the potential to open up in these areas, this work aims to carry out more in-depth research into stereoscopic techniques and their applications. Thus, the primary objective of this project is to compare the advantages and disadvantages of the stereoscopic solutions discussed, taking into account qualitative and quantitative factors, generating discussions and presenting comparative information that can help in the selection of which solution to expand or implement in future applications. The secondary objectives are to hold discussions and propose trends in order to indicate the most appropriate solutions to be implemented in a system that exploits telepresence using video avatars and in the application of stereoscopy in the area of games and graphics engines.

An essential contribution of the work and means of achieving the primary objective is the proposal of a methodology for measuring and comparing the performance of stereoscopic techniques in different scenarios. Extensive theoretical research into techniques that enable stereoscopic visualisation is carried out, which provides the conceptual foundation needed to design and implement the test scenarios and select which stereoscopic techniques to evaluate. Having defined this theoretical process on stereoscopy, the experimental part remains to be realised. The path chosen is the creation of a test system, which allows both modules with new stereoscopic solutions and new test scenarios to be coupled.

Another contribution that is more related to the secondary objectives is the discussion of the use of stereoscopy in the area of games. The discussion aims not only to present technical limitations and solutions, but also in terms of game design. Stereoscopy provides game designers with new variables that can be exploited in game mechanics to provide new experiences for players. From a technical point of view, one of the main targets of the argument is EnJine, with a focus on incorporating a stereoscopy module into this game engine. Although the implementation of the module is not within the scope of the work, relevant points to guide a developer implementing this module are detailed. In addition, limitations of both Java 3D and EnJine that restrict or prevent implementation are described.

1.1 Justification

Stereoscopy has been a topic of curiosity for researchers since at least the mid-19th century (FEHN, 2005) and its study and development has largely been driven by initiatives aimed at commercial exploitation. An indication of this commercial focus can be seen in the number of patents and exclusive products from companies working in this area (ROBERTS, 2003; APO, 2008).

Due to the large number of isolated commercial initiatives aimed at reaching the public, stereoscopy solutions have become superficially known, but rarely explored in depth, especially in relation to the classification and development of solutions. One effect of this diluted knowledge is the large number of personal and commercial websites dealing with the subject of stereoscopy, providing tips and products. In view of these points, stereoscopy is a subject that still requires consolidation efforts, so that it can reach a stage of maturity that allows it to be widely used and not just for restricted use or as an exotic differentiator.

As presented in the previous chapter, one field in which stereoscopy has potential is in virtual and augmented reality, as it gives the user a greater sense of telepresence. Especially for the Video-Avatar project developed at INTERLAB, stereoscopy is a natural requirement, as it gives the participant a more realistic experience. Figure 1 shows some images of the concept used in AVTC, one of the products of the Video-Avatar research line.

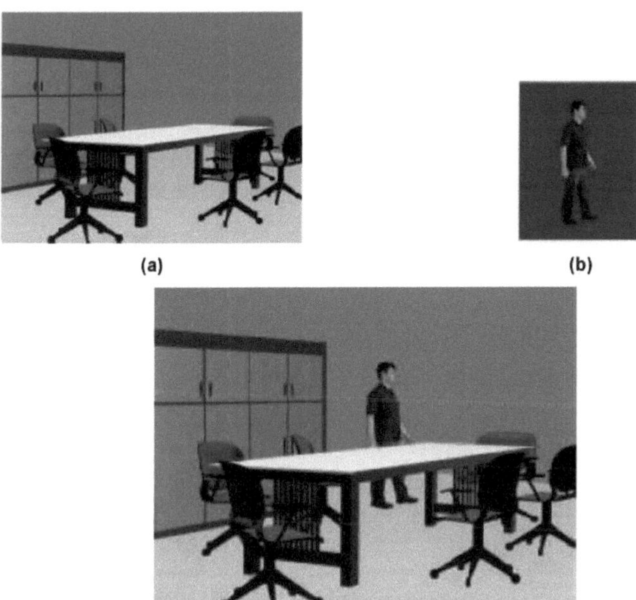

Figure 1 - AVTC conceptual images. Virtual environment (a), Video of the person (b) and the result of the composition (c) (SISCOUTTO, 2003).

A further incentive in this area of research is the increasing investment by large companies involved in computer graphics and visual communication. NVidia, one of the major graphics card manufacturers, has resumed work on its software that allows stereoscopic visualisation in anaglyphs, shutter glasses, among others, and has launched a new version of the programme that supports some monitor models with polarisation stereoscopy (NVIDIA, 2008).

From the consumer's point of view, the cost of obtaining the equipment needed for stereoscopic visualisation is falling and the advance of auto-stereoscopy makes it possible to visualise three-dimensional scenes without the need for special glasses (WEINAND, 2005b).

The cinema industry is investing more and more in 3D films and it's no coincidence that cinemas are increasingly adopting the digital system, which supports stereoscopic projection (SCREENDIGEST, 2010).

1.2 Virtual Reality and Games

Virtual Reality (VR) is typically seen as a new type of media and defined through equipment such as computers, Head Mounted Displays, headsets and gloves with movement sensors (STEUER, 1992).

However, this is a weak definition of VR, as it links the theoretical concept to the technological solutions used. A better definition of VR could be based on the concept of telepresence described by Steuer (1992), in order to isolate the concept of VR from the technology, which is always evolving. Thus, a possible definition of Virtual Reality, which is the one adopted in this text, is that of a real or virtual environment in which the user experiences tele-presence.

At the same time as VR research, an area that has also benefited from the evolution of computers is the electronic games industry, aimed primarily at the market. Games went from monochrome 2D animations to a major boost in the development of virtual environments in three dimensions. From the 1990s onwards, the video games community began to occupy spaces that belonged to the VR community. Research into VR now takes into account what happens in the area of video games, focusing not only on entertainment, but also on training, simulation and education (ZYDA, 2005).

A feature of games that is of great value in other areas that create virtual environments are game engines. These engines accumulate the main functionalities needed to generate a virtual environment, such as displaying 3D models and accessing input and output devices. This integration of different functionalities in the same module allows the component to be reused in different projects and applications (FRITSCH; KADA, 2004).

1.3 Augmented Reality

The concept of Augmented Reality covers a region within an imaginary line, where the extremes are the Real World and Virtual Reality, as defined by Milgram (1994).

The Real World is nothing more than the environment we live in, where all the objects are physical, while Virtual Reality represents a completely computer-generated environment, with only virtual elements. To be classed as Augmented Reality, an environment must consist predominantly of real elements, with some additional virtual ones. Figure 2 shows the representation of this Real-Virtual line.

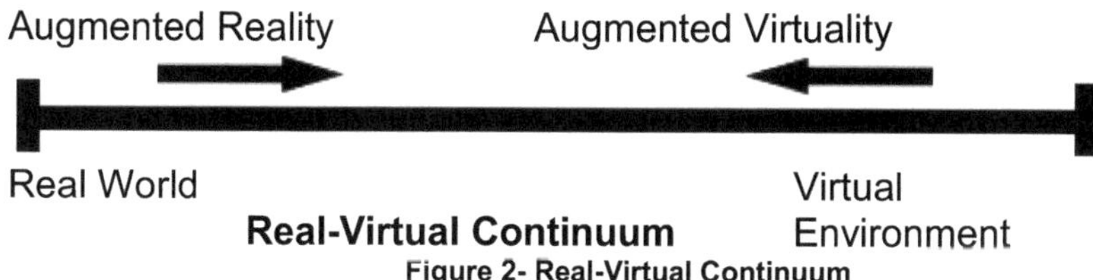

Figure 2- Real-Virtual Continuum

Another region, defined as being closer to the virtual extreme than AR, is Augmented Virtuality, in which virtual elements predominate in the environment, with some real parts added.

As this is just a conceptual definition of Augmented Reality, a more technical definition was needed, which would allow us to classify more objectively whether a system is AR or not.

Azuma (1997) proposes that for a system to be AR, it must exhibit three characteristics:

- Combining real and virtual
- Be interactive in real-time
- Be registered in 3D

The first characteristic is trivial, resulting from the very nature of AR being between the Real World and Virtual Reality. The second characteristic allows us to isolate false applications that may appear to be AR at first glance, for example the films "Who Framed Roger Rabbit" from 1988 and "Space Jam" from 1996, in which cartoons interact with real characters. One point they don't meet to be considered AR is the requirement to be interactive in real time, as viewers can't manipulate the elements of the film. Lastly, 3D recording, which also filters out false

AR candidates, such as a projection on a screen, although interactive and combining real and virtual, are not recorded in three dimensions.

An example of a project that aims to fully utilise Augmented Reality is INTERLAB's Video Avatar, which was presented earlier in section 1.1. The aim is a homogeneous integration between the virtual and the real, following the definitions mentioned by Azuma. The user will be positioned within the 3D virtual environment represented by their video avatar and will be able to interact with the environment through the avatar, moving around and interacting with objects.

Due to the potential for applications and innovation, Augmented Reality is used in various areas of knowledge, including medicine, education and military training.

One AR niche that has gained its favour is Spatial Augmented Reality. In this category of AR, images are projected onto real objects in order to provide an environment with virtual elements for an observer (RASKAR et al., 1998). A major advantage of this method is the simplicity of visualisation, as the virtual elements exist in the real world as projections, there is no need to use equipment such as Head Mounted Displays.

One project developed at INTERLAB that uses the concept of Spatial Augmented Reality is the Robot ARena framework (CALIFE et al., 2007). This project consists of a real robot built with Lego Mindstorms pieces that is positioned on a table, where virtual elements such as barriers and collectable objects are projected. The robot's position and orientation are captured by a camera positioned a few metres above the table

One of the applications created on this framework is Footbot AR, a football game between robots, in which one of the robots is real and the other virtual, generated in Spatial AR. Another game created on this platform is TanSpace, which uses the concept of a Tangible Interface. To interact with the projected virtual elements, the player manipulates a marker with their hands that must be aligned with a virtual crosshair to destroy enemy ships.

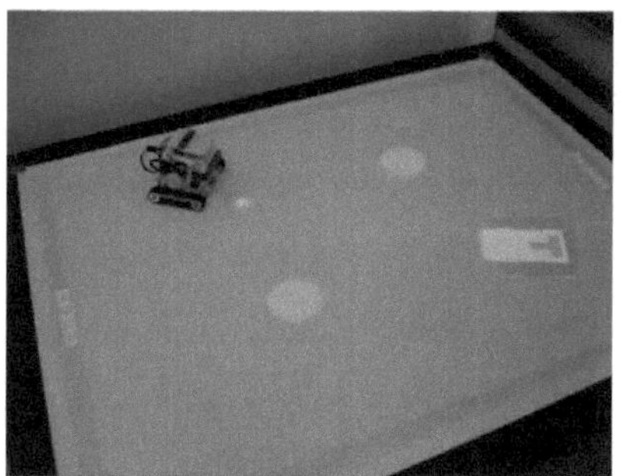
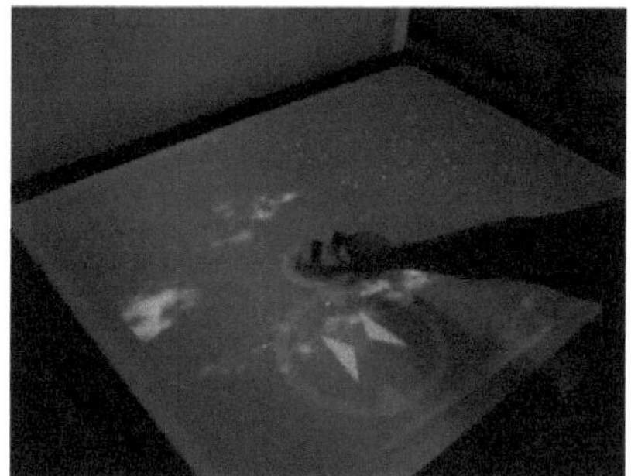

Figure 3 - Images of FootBot AR (left) and TanSpace (right) by Daniel Calife.

1.4 Structure of the work

This work is presented in eight topics, which can be divided into two groups of four, the first consisting of this Introduction, item 2 Stereoscopy, which describes the theoretical information obtained in the literature research, item 3 Methodology, which reports on the procedure used to select the stereoscopic techniques and the design and execution of the tests, item 4 Experiments, which presents the process of creating the test system and carrying out the battery of tests. These topics
represent the conceptualisation and execution of the work. The other topics make up the analysis, discussion and conclusion stages of the research, with item 5 Analysis reporting the theoretical and experimental analysis of the stereoscopic solutions, item 6 Application in AR presenting discussions on stereoscopy and AR, item 7 Application in Games containing a discussion on the relationship between stereoscopy and games, and item 8 Conclusions describing the closing of the text with the conclusions of the research and proposals for future work.

CHAPTER 2

STEREOSCOPY

The term stereoscopy comes from two Greek terms that stand for "vision" and "solid" and the origins of this area of knowledge can be traced back to at least 1838. This is the date on which the English researcher Charles Wheatstone developed a device that enabled an observer to visualise a three-dimensional scene from images with different perspectives (FEHN, 2005). This initial version of the stereoscope was improved by David Brewster in 1844 and by the end of the 19th century it was present in many homes in Europe and the United States. Since then, many researchers have endeavoured to discover how the brain identifies the depth of elements in a scene. Bela Julesz took an important step in this direction in the 1960s, when she demonstrated using random dot stereograms (detailed in section 2.2.4) that the brain does not need monocular vision cues to make this calculation (BELTRÃO, 2007). This monocular information, such as perspective and shadow, is presented in section 2.1 below.

2.1 The "Solid Vision"

The main characteristic that enables humans to have solid vision, i.e. the ability to identify the depth of elements in a scene, is bi-ocular vision. Through the disparity in the image seen by each eye due to the difference in perspective, the brain is able to determine the depth of what is being seen (HOLLIMAN, 2002).

From an evolutionary point of view, stereoscopy was a feature that arose in parallel in at least two groups: mammals and birds. Although only a few birds have stereoscopic vision, practically all mammals have this feature, albeit in a rudimentary form. This leads to the conclusion that stereoscopic vision was a characteristic present in the first mammals and was of great value to the growth and success of this group (PETTIGREW, 1986).

Figure 4 illustrates the various parts of the human eye. The pupil is located between the cornea and the lens and is the entry point for light. The retina is located in the inner part of the eye, covering most of the inner surface.

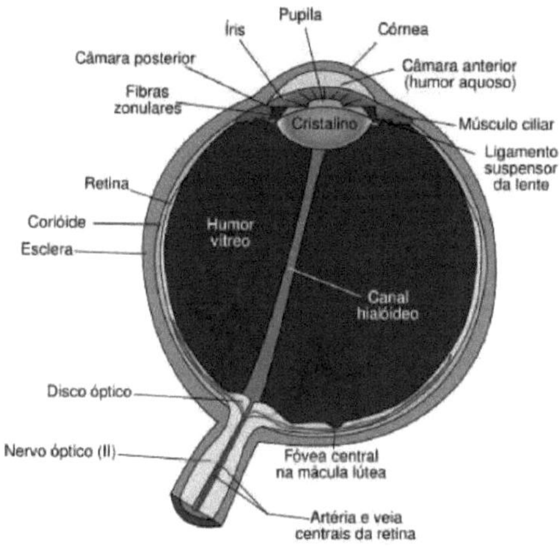

Figure 4 - Anatomy of the human eye (WIKIPEDIA, 2008).

Stereoscopic vision has a great advantage over monocular vision, as it allows the individual to obtain a range of additional information about the environment:

- **Relative depth recognition** - The observer can identify the spatial relationship between objects instantly;
- **Spatial localisation** - Allows the brain to focus on objects positioned at a certain distance, ignoring other depths;
- **Camouflage identification** - The ability to recognise camouflaged objects in a scene without having to move. This is probably one of the most important characteristics that favoured stereoscopic vision during the evolutionary process;
- **Perception of the surface of materials** - The difference in the image seen by each eye is one of the characteristics responsible for identifying lustre in metals and precious stones;
- **Curved surface recognition** - Biocular vision makes it easier to recognise curved surfaces.

Some of this information, especially the identification of camouflage, can only be recognised through monocular vision if the observer moves, generating what is known as **movement parallax**. In animals that do not have stereoscopic vision, or are unable to use it, motion parallax is important information for discovering the position of objects. Collett (2004) gives an example of an insect using motion parallax to determine the distance to prey.

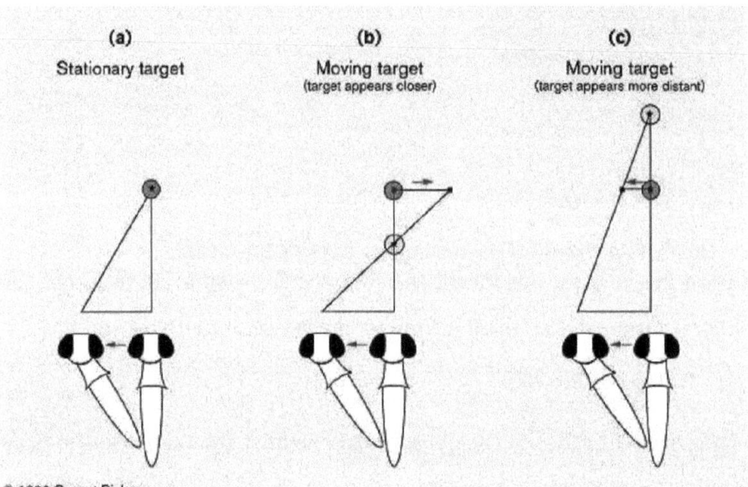

Figure 5 - From left to right: (a) the insect moves its head to determine the distance from the object; (b) the object is simultaneously moved in the opposite direction, appearing to be closer; (c) the object is moved in the same direction, appearing to be further away (COLLETT, 2004).

Holliman (2002) demonstrates mathematically that biocular vision has distance restrictions. In the case of humans, stereoscopic vision is efficient from a few centimetres from the eyes, up to a few tens of metres away. Beyond a few hundred metres, it becomes very inefficient, allowing only a rough estimate of distance.

Although stereoscopy provides the primary depth information to the human brain, other characteristics of the scene also help to visualise the correct positioning of elements in space Siscoutto et al. (2006) and Holliman (2002) present the following effects:

2.1.1 Perspective

Because of the anatomy of the human eye and the way light enters through the pupil hole to be detected by the retina, real-world objects that are closer to the observer appear larger than more distant objects, even though both are the same physical size.

A consequence of the perspective effect can be seen in Figure 6. Both parallelepipeds are the same size when compared to a ruler. However, our brain, interpreting that this is the image of a long corridor, leads to the illusion that the rightmost parallelepiped is smaller than the centre one.

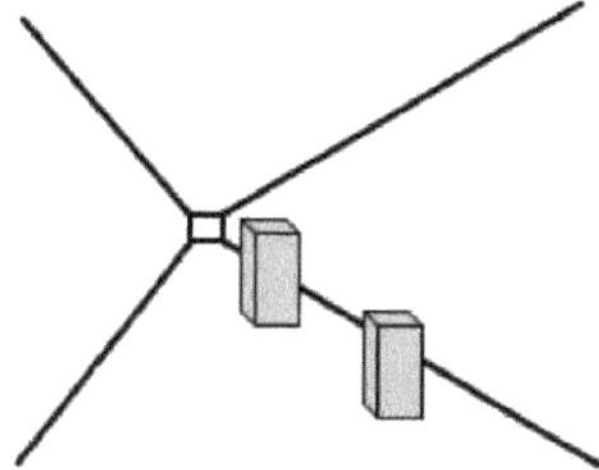

Figure 6 - Example of perspective effect (based on SISCOUTTO et al., 2006)

2.1.2 Lighting and shade

The presence of lighting adds information to the scene that allows the brain to recognise both the surface of objects and their location more accurately. The differences in colour on the surface of objects caused by lighting create important clues as to their real shape and the existence of an object's shadow on other surfaces acts as a reference for its spatial position.

Figure 7 below shows three different situations to illustrate the effect of lighting. In the first, without lighting, the objects look like a monochrome circle and hexagon. In the second, the shadow due to the lighting generates colour differences on the surface of the objects, which makes it possible to identify them as a sphere and a cube. Finally, the shadow on a flat reference surface indicates that the cube is leaning against it and the sphere is floating on the surface.

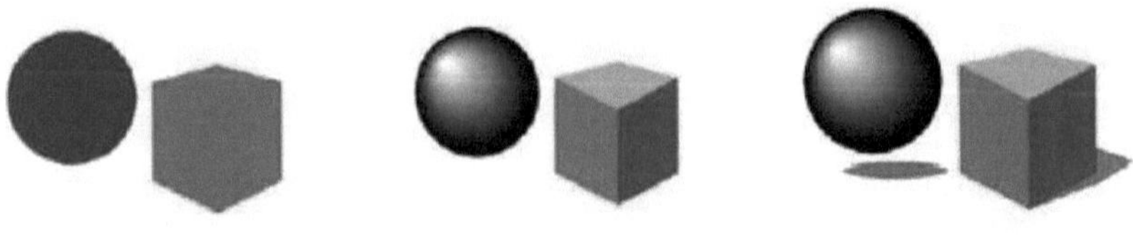

(a) (b) (c)

Figure 7 - Example of the lighting effect. (a) No lighting. (b) Lighting with shadows on the surface of the object itself. (c) Same as (b) adding shadows on other surfaces. (SISCOUTTO et al., 2006)

2.1.3 Occlusion

The occlusion effect occurs according to the position of objects in space. As light propagates in a straight line, objects that are closer to the observer hide more distant objects that are in the same line of sight. Total occlusion of an object prevents it from being seen, but partial occlusion provides the brain with important information to recognise the relative position of objects.

Figure 8 shows a sphere and a cube in two different situations, in which the occlusion effect defines how we interpret the relative position between them. In the image on the left, the cube partially covers the sphere, leading the brain to interpret that the cube is in front of the sphere. In the image on the right, the opposite is true: the sphere appears to be in front of the cube.

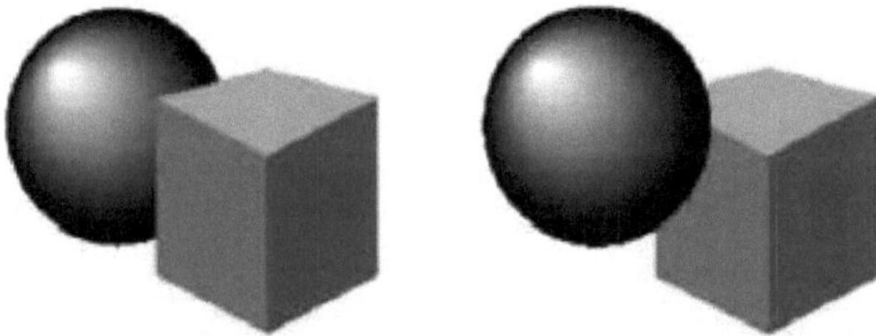

Figure 8 - Example of partial occlusion (SISCOUTTO et al., 2006).

2.1.4 Texture Gradient

This effect occurs on large surfaces with a similar texture throughout, such as a lawn. As the distance from the viewer increases, the details become very small and unidentifiable. Figure 9 shows an image of a lawn, in which the distinction that the lower part is closer to the viewer than the upper part is recognisable due to the texture gradient.

Figure 9 - Grass image illustrating the effect of the texture gradient (PHOENIX TROPICALS, 2008).

2.1.5 Middle Distortion

Holliman (2002) uses the term air distortion to describe this effect. However, I use the broader term medium distortion because it is an effect that is not restricted to situations in which the observer sees light that has propagated through the air. Light rays are susceptible to distortions on the way between objects and the observer's eyes, caused by various reasons such as fog, dust or rain. The greater the distance the light has to travel, the less detailed the contours become and the colours lose their saturation. In the following pictures, even without considering other visual effects, you can recognise the trees behind by the whitish colour and identify the part of the wrecked ship that is further away by the bluish colour.

Figure 10 - Example of distortion of the medium caused by mist suspended in the air (O'BRIEN, 2008).

Figure 11 - Example of distortion when the medium is water in an image taken by a diver (DAVY JONES LOCKER, 2007).

2.2 Stereoscopy solutions

2.2.1 Stereoscope

This was the equipment created by Wheatstone to enable an observer to see a scene in three dimensions. The stereoscope is based on a scheme of lenses and mirrors that direct the correct images to the left and right eye following principles of optics. The equipment also allows the physical configuration between the eyes and between the observer and the images to be adjusted as required (SISCOUTTO et al., 2006).

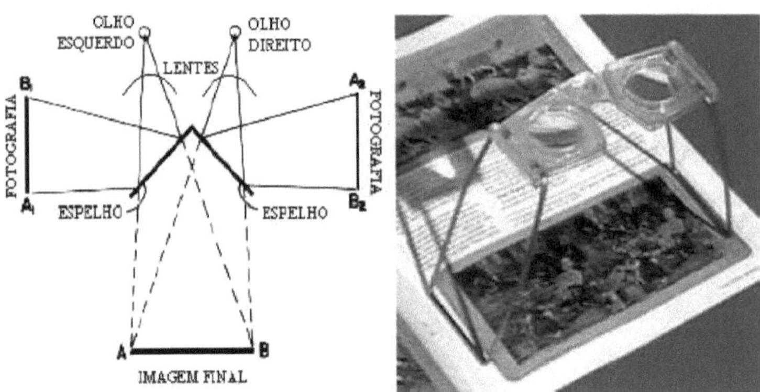

Figure 12 - Basic diagram of a stereoscope and an example of this equipment (SISCOUTTO et al., 2006).

2.2.2 Stereo Slides

Another piece of equipment used to achieve a stereoscopic effect is a stereo slide viewer. This equipment consists of a small box resembling binoculars, in which a pair of stereo slides is placed on one side and the opposite side has two holes for the viewer to position their eyes.

The Rocky Mountain Memories website (2001) teaches you how to create your own stereo slides for various viewer standards. One point that is emphasised on the website is the Stereo Window effect, which occurs due to the limited field of view that a slide contains. For the montage to be coherent, no element of the image at the edge of the slide needs to be positioned outside the plane of the slide when viewed in 3D.

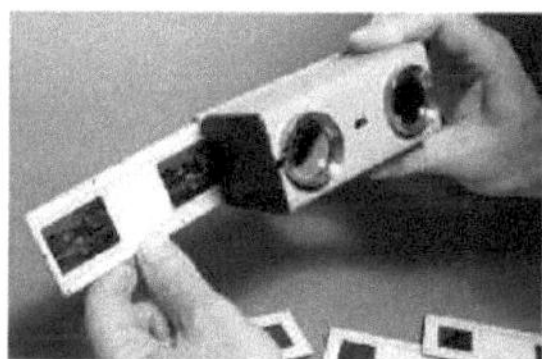

Figure 13 - Examples of stereo slide visualisers (REEL 3D, 2008).

2.2.3 Stereo Pair

A stereo pair is simply the presentation of two images that form a stereoscopic pair side by side, which can be observed without the aid of any equipment. In order for the observer to be able to see the stereoscopic effect with the naked eye, it is also necessary to use one of the visualisation techniques presented below. Figure 14 shows two examples of a stereo pair, one in parallel view and the other in cross-view. An interesting effect that can be noticed when viewing the images is that when you position your eyes to view one of the pairs, you can also see the other pair in stereo, but without the same depth clarity. The conclusion is that even by inverting the information from the eyes, the brain is able to interpret the images as a coherent pair and form the three-dimensional scene.

Parallel View

To use the parallel vision technique, the images need to be positioned side by side, with the left image on the left and the right image on the right. The stereoscopic effect is achieved if the observer looks at the images by darting their eyes, but mastering the technique can be difficult for a beginner (COLLIER, 1993). When successful, three images are visualised, with the central image being a stereoscopic version of the scene contained in the other two images. This technique has a human limit to the size of the images, as the eyes cannot diverge many degrees. A tip for mastering this visualisation technique is to look through the images as if you were observing a distant point, relaxing your eyes until the third central image comes into focus.

Cross vision

In the cross-eye technique, the images must be positioned in reverse, i.e. the left image on the right and the right image on the left. Unlike parallel vision, to achieve the stereoscopic effect in cross-eye vision the viewer's eyes must converge. As the eyes can converge at a greater angle, this technique makes it possible to visualise much larger images (COLLIER, 1993). To learn this technique, the tip is to place a finger between your eyes and focus your vision on the finger with the images in the background. By moving your finger towards the images, there will be a point at which all three images will be visualised; this is the angle your eyes must be at to see the central image in 3D.

(a)

(b)

Figure 14 - The upper stereo pair (a) is set up for parallel vision and the lower one (b) for cross vision (KLEIN, 2008)

2.2.4 Stereogram

Stereograms are simple 2D images that contain the stereoscopic information hidden in them. Two classic forms of stereogram are random dot and texture stereograms, both of which can be visualised using the same technique applied to viewing a stereo pair. The principle of the stereogram is to combine the view of the left and right eye into a single image. However, in order to combine the two images, it is necessary to readjust the colours of the pixels in order to maintain coherence between a pixel of a point seen in the left image and the pixel of the same point seen in the right image (MINH et al., 2002). Because of this restriction, the final result of the stereogram ends up being an image with random points or some kind of texture.

Figure 15 - On the left a stereogram of random points with the text "3D" and on the right a stereogram of texture with a question mark (BEENE, 2008; BWH VENTURES, 2008).

One aspect of the stereogram technique is animated stereograms, which contain a sequence of images rather than a single stereo image of the scene. Some examples of animated stereograms can be found on the internet, on Beene's website (2008) there are animations of a fan and a pendulum.

There are also efforts to use stereograms in computer games. A pioneering game in the stereo vision approach is Magic Carpet (Bullfrog - Electronic arts, 1994) which implements stereoscopic rendering in the form of stereograms as well as anaglyphs.

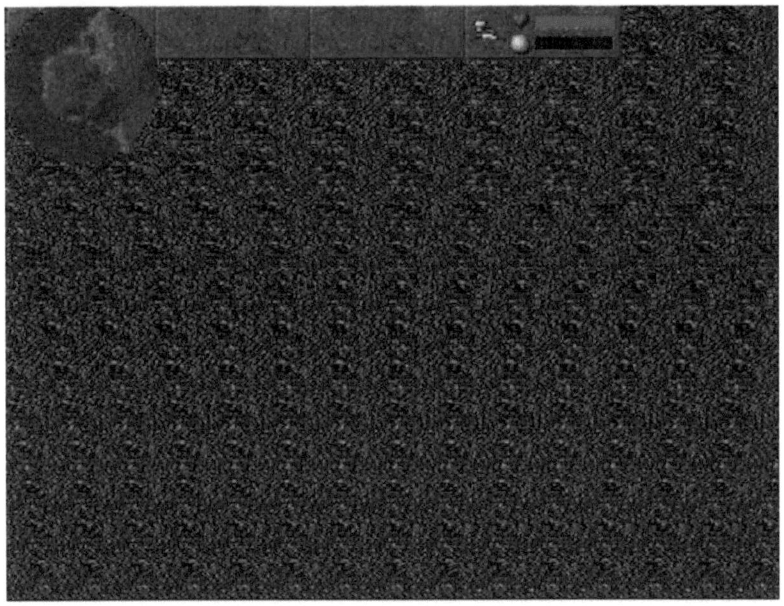

Figure 16 - Stereogram image of the Magic Carpet game. The scene is an area view of a hill with sparse vegetation and a "TF"-shaped structure in the centre.

2.2.5 Pulfrich effect

Stereoscopy using the Pulfrich effect is based on the characteristic of human vision to perceive less intense light more slowly. To create the three-dimensional effect, the viewer has to use a filter that makes vision darker for one eye, so when watching an animation the eyes have a different speed of perception, leading to a slight displacement between the images seen by each eye, which creates the sensation of depth. The problems with this method are the lack of greater depth control and the limitation that it only works with moving scenes (SISCOUTTO et al., 2006).

Figure 17 - On the left, glasses with a filter to create the Pulfrich effect and on the right, frames from an animation, while the eye without the filter perceives the current frame 2, the eye with the filter still sees frame 1 (SISCOUTTO et al., 2006).

One commercial application of this technique was the creation of VHS videos that came with glasses based on the Pulfrich effect. One example is the 1998 video "A Walk Through the Roses of Reynolda Gardens" produced by Dave Combs (2008).

2.2.6 ChromaDepth

This form of stereoscopy is based on the use of colour to define the depth of elements. The lenses of glasses with ChromaDepth technology have the property of deflecting light according to its colour, providing the disparity between the vision of the left and right eye necessary for the three-dimensional effect. In this technique, red-coloured elements are seen closer to the observer and the further the colour of the object moves in the visual spectrum towards blue, it appears to be further away (SISCOUTTO et al., 2006). This effect is illustrated in Figure 18.

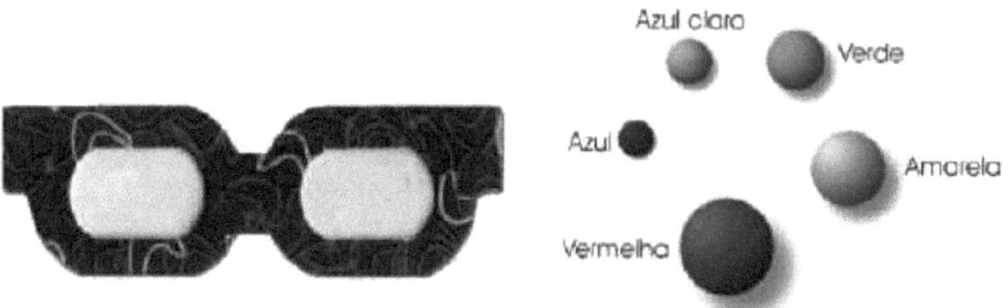

Figure 18 - Glasses with ChromaDepth filter and difference in perceived distance for different colours (SISCOUTTO et al., 2006).

This solution presents the direct problem of making it impossible to use colour naturally, as the colours of the different points in the scene contain the additional information of their depth. The glasses are sold in two versions: Standard (STD) for print media and High Definition (HD) for electronic media (APO, 2008).

2.2.7 Anaglyph

The anaglyph technique exploits the division of the image into its colour spectrum, allocating a range of colours to the left eye and another complementary range to the right eye. To view an anaglyph image, the viewer must wear special glasses with filters that allow only the colours allocated to each eye to pass through. The most common filter in anaglyphs is to use red in one of the lenses and blue or cyan in the other. Thus, the eye viewing through the red lens only perceives light in the red range and the eye with the blue/cyan filter only perceives light in the respective range (SISCOUTTO et al., 2006).

Figure 19 - Classic anaglyph with a grey-scale image when viewed with the naked eye and anaglyph viewing glasses with red and blue filters (3D EXPERIENCE, 2006; RAINBOW SYMPHONY, 2000).

A variant of the classic technique for generating anaglyphs are the so-called coloured anaglyphs or colourglyphs. The main difference with coloured anaglyphs is that they display much of the image's real colour when viewed with the naked eye, but when viewed in 3D through filters, much of the colour information is lost.

Figure 20 - Coloured anaglyph, also called colorglyph (BURDER, 2005).

The anaglyph solution was conceived in the photographic field, and its invention is attributed to the inventors Joseph D'Almeida and Louis Ducos du Hauron in the 1850s (LPI, 1999). In 1858, D'Almeida gave a 3D slide show to an audience wearing glasses with red and blue filters. In addition to anaglyphs, Hauron also developed the trichrome method in 1869, used in colour photography, and created printed anaglyphs (BRITANNICA, 2008). The first anaglyph film was shown in 1893, created by William Friese-Green.

2.2.8 ColourCode 3-D

This solution, in essence, uses the same idea as the anaglyph technique: separating the image that each eye receives by isolating one band of the spectrum for the left eye and another for the right eye. The difference between the two techniques lies precisely in the range of the spectrum determined for each eye. Instead of trying to evenly divide the range of colours that each eye receives, ColorCode filters separate them in an unbalanced way, leaving a small portion of frequencies for one eye and all the remaining frequencies for the other. It's as if one eye receives all the colour information in the image, with a shift in the observed colour palette, and the other eye complements it with depth information, as if it were seeing in greyscale (COLORCODE, 2008).

Figure 21 shows an example of a ColorCode image. Seen with the naked eye, you can see that the image can be divided into a main coloured part, in yellowish tones, and a blue shadow, which is the part with the depth information. The same figure also shows a pair of glasses with ColorCode filters that have amber and blue colours.

Figure 21 - 3-D ColorCode image and glasses with ColorCode filters (COLORCODE, 2008).

The ColorCode 3-D patent was published in 2004 and was signed by three Danish researchers (US PATENT 6687003). The patent shows an image of the response of ColorCode filters in transparency level for a certain wavelength of the electromagnetic spectrum. Figure 22 shows two graphs covering the wavelength range of visible colours. In the first graph, the curves represent the sensitivity of human eyes to a given wavelength, considering that the human visual system is trichromatic and has three categories of light-sensitive cones, with different points of maximum sensitivity (DEEB; MOTULSKY 2005):

- Blue Cones ~420nm *(short wavelength);*

c Green Cones ~530nm *(middle wavelength);*

- Red Cones ~560nm *(long wavelength).*

The second graph shows the response of the ColorCode filters. Based on the graph, it can be concluded that one of the filters allows wavelengths that sensitise blue cones to pass through, while the other filter is complementary, allowing light that sensitises green and red cones.

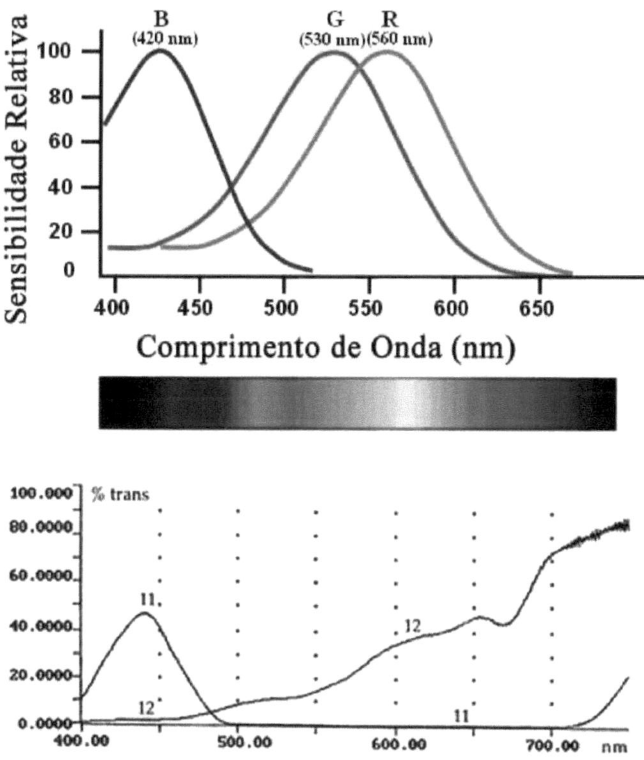

Figure 22 - Graphs showing the sensitivity of the human eye (top) and the response of ColorCode filters (bottom) to different wavelengths of visible light (based on DEEB; MOTULSKY 2005; US PATENT 6687003).

2.2.9 Polarisation of Light

Visible light constitutes a band of the electromagnetic spectrum and is propagated by electromagnetic waves that vibrate in an arbitrary way in all directions perpendicular to the direction of travel. The process of polarising light means that these directions of vibration are no longer arbitrary, making the light vibrate in an organised way. Light is normally polarised in three patterns: linear, elliptical and circular, with circular polarisation being a special case of elliptical polarisation (NAVE, 2006). Figure 23 shows these three polarisation formats, which can be viewed graphically on the

WebPhysics website (WOLFGANG; DANCY 2008).

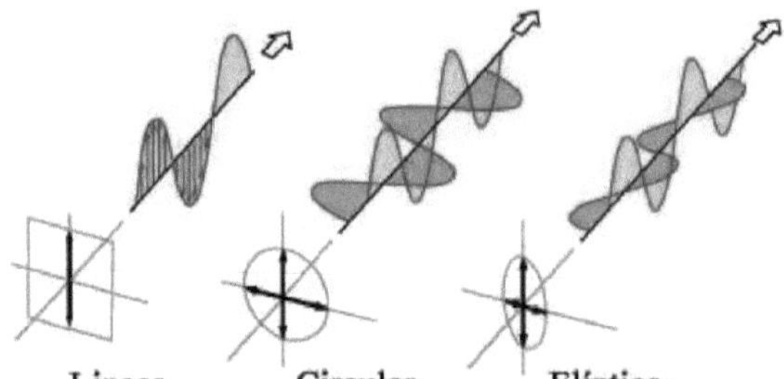

Figure 23 - Light polarisation patterns (based on NAVE, 2006)

This concept of polarisation can be used in the field of stereoscopy as a technique to separate the information received by each eye and thus make stereoscopic viewing possible. To do this, the left and right images need to be superimposed, but with different polarisations, so that an observer wearing glasses with the appropriate polarising filters receives the correct image for each eye. Furthermore, in order to obtain the expected result, it is important that the filters used for each eye produce waves that are orthogonal to each other. In the case of linear polarisation, this is achieved by using one filter horizontally and the other vertically, while in circular and elliptical polarisation, one filter must be anti-clockwise (also called left) and the other clockwise (also called right). A filter is classified as clockwise or anti-clockwise according to the phase difference in the vertical and horizontal components, if the vertical is delayed in relation to the horizontal (as in Figure 23) the filter is clockwise and otherwise anti-clockwise.

A common practice for applying polarisation in stereoscopy is to use two projectors aligned with orthogonal filters and project the stereoscopic pair onto a screen that maintains the polarisation of the light. Observers then wear glasses with filters compatible with those placed on the projectors, which allows each eye to receive only the image produced by one of the projectors, making stereoscopic visualisation possible (SISCOUTTO et al., 2006).

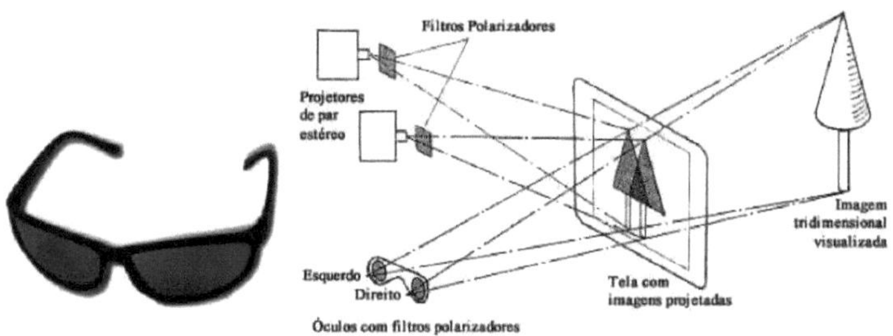

Figure 24 - Glasses with polarised filters and scheme for stereoscopic viewing using polarised light (AMAZON, 2008; based on WAACK, 2004).

One final note is the restriction on the observer's movement when using linear polarisation. As one filter is horizontal and the other vertical, if the observer is not aligned with the screen, the stereoscopic effect will not be observed. In the case of circular polarisation, this problem is eliminated, making this type of polarisation more flexible and robust in less controlled environments.

2.2.10 Shutter *Glasses (LCS/Shutter Glasses)*

In this stereoscopic solution, the left and right images appear alternately to the observer and equipment is needed to block the observer's vision so that they can see the images through the correct eye. This equipment is usually shutter glasses that use liquid crystal lenses and block the viewer's vision in sync with the images being displayed (ENGDAHL, 1999). This technology has been widely developed commercially and there are several patents registered in this area (STARKS, 2008).

This technology has been exploited both for 3D visualisation on computers and televisions. Engdahl (1999) lists several of the different standards that have been created to achieve stereoscopic perception through shutter glasses.

Even considering only computer solutions, many different formats have been devised (WOODS, 2005), the main ones being:

Page Flipping - Produces the best results among the formats, it consists of switching directly from one image to the next directly in the video memory;

- **Interlaced** - Probably the simplest method that uses the same scheme as the PAL and NTSC standards of alternately updating even and odd lines. The images from one eye are displayed on the even lines and from the other on the odd ones;
- **Sync Doubling** - This works by dividing the image into two halves and associating one eye with the upper half and the other with the lower half. An external vertical synchronisation signal is always generated at the vertical midpoint, so that each half is displayed as a complete image of the monitor;
- **Line Blanking** - Used when interlaced images are shown on a non-interlaced monitor. An external device cleans up the information from the lines not updated in that frame so as not to mix up the images from the two eyes.

In addition to the different formats for generating images, there are different standards for connecting the glasses to the computer and making it possible to synchronise the generation of the images with the blocking of the viewer's eyes. Some models are connected to a special output on the video card, which signals the vertical synchronisation pulses to the glasses. Other models are connected directly to the VGA output, routing the video signal to the monitor and the vertical synchronisation to the glasses. There are also solutions that offer wireless shutter glasses, in which synchronisation is done by infrared transmission.

Figure 25 - Obturator glasses in wired and wireless versions (SISCOUTTO et al., 2006).

This technology has two main problems: crosstalk and flicker (ENGDAHL, 1999). The problem of crosstalk consists of the system's inability to completely isolate the images to be shown to the left eye from those intended for the right eye. As a result, the eyes end up seeing "ghosts" of the images intended for the other eye. In the case of flicker, the problem is caused by the system's lack of speed in switching between images, which makes the stereoscopic result inconsistent for the viewer. As the refresh rate for each eye is divided by two, the system's frequency must be at least 100Hz to 120Hz to achieve a refresh rate of 50Hz to 60Hz, which is the standard used in television.

2.2.11 Head Mounted Display

Equipment known as Head Mounted Displays (HMDs) are so closely linked to the development of virtual and augmented reality that there are authors who associate their use as a requirement for creating such environments. The great advantage of HMDs in terms of stereoscopy is the complete isolation between the image generated for the left eye and that generated for the right eye, allowing for a completely crosstalk-free visualisation (AZUMA, 1997). HMDs are classified into two basic categories: optical and video.

Video HMDs

In this type of HMD, the viewer only sees the content that is shown via video on the device's displays. The video may contain scenes from the real world, but these scenes are first captured by a camera and then passed on to the viewer. In an augmented reality system, this real-world video is augmented with virtual elements, which need to have their position correctly registered in relation to the real objects.

An HMD of this type can be seen in Figure 26, along with an image of an augmented reality system. The major advantages of this technology are:

- Flexibility to compose images of real and virtual content;
- Wide field of vision because optical distortions can be corrected digitally;
- Synchronised updating of real and virtual content;
- The digitally captured scene can be used to help register the virtual elements.

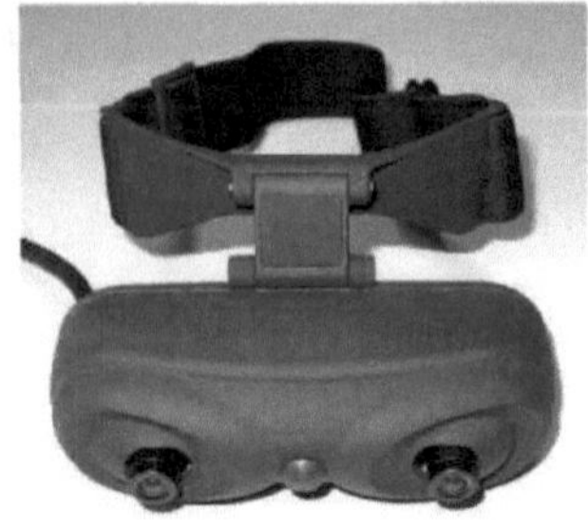

Figure 26 - Image resulting from a system using a video HMD and the AR-Vision-3D HMD with see-t/7roug/7 *video* technology and two coupled cameras (KONDO et al., 2007; INITION, 2008).

A major safety problem for observers using such equipment is their dependence on the device to visualise the real world around them. If there is a fault and the video of the real world is no longer displayed, the observer will be "blind" until they remove the equipment.

Optical HMDs

The key part of this equipment are the semi-transparent lenses in which the natural view of the real world is combined with the virtual content generated by the system. Unlike video HMDs, any manipulation of real-world images that the viewer receives can only be manipulated through optical schemes. In the event of a system failure, the user will continue to see the real world, but the intensity of the light received will be lower due to the semi-transparent nature of the lenses.

Figure 27 shows an image resulting from such a system and an example of an optical HMD. The main advantages of this type of HMD are presented below:

- Simplicity and lower costs;
- The resolution of real images is limited only by the human optical system;
- Safer than video HMDs. In the event of failure, the viewer can still see the real world, just as if they were wearing sunglasses;
- The point of view of the observed images is exactly the position of the eyes. In the case of video HMDs, the images are relative to the position of the cameras.

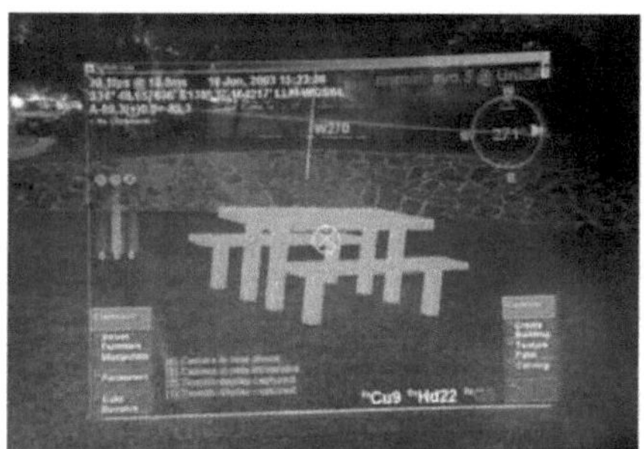
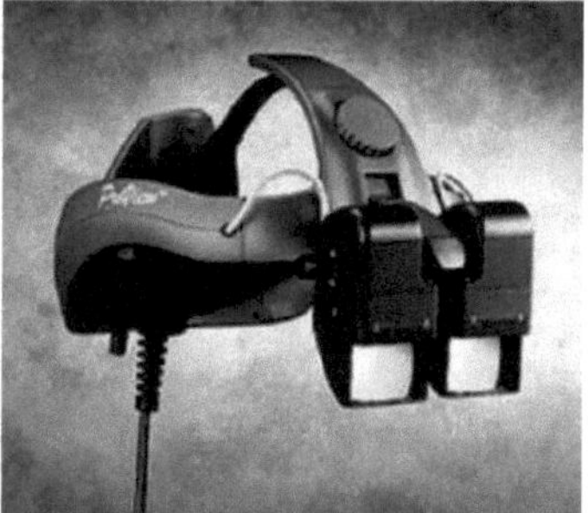

Figure 27 - Image resulting from system using optical HMD and HMD ProView XL40/50 with optical see-through technology (PIEKARSKI, 2004; SILVA, 2003).

1.1.12 Autostereoscopy

This category of stereoscopic solutions encompasses many different products and technologies. Depending on the definition of autostereoscopy used, this group can cover any type of solution that allows you to observe with a sense of depth with the naked eye. However, in the scope of this work, an autostereoscopic solution will be considered to be one that fulfils the following requirements:

1) It allows stereoscopic visualisation without the observer having to use any optical equipment to separate the left and right channels;

2) It doesn't require any visualisation techniques on the part of the observer;

The first restriction comes from the very nature of the term autostereoscopy, which reflects stereoscopic visualisation without the need for optical equipment. The second requirement aims to restrict the concept of autostereoscopy a little further, so that any observer with normal vision and no prior knowledge of the solution can visualise the scene with a sense of depth, thus excluding, for example, stereogram and stereo pair solutions.

Considering these requirements, we present some of the solutions in this area, which has been the subject of intense research and has been widely cited at the "Stereoscopic Displays and Applications" conference and by companies that manufacture this type of product.

Parallax barriers

The first type of autostereoscopic solution were parallax barriers, which consisted of dividing two or more images into vertical strips and positioning them behind opaque vertical barriers of the same frequency, so that an observer would see different images depending on their point of view (ROBERTS, 2003). The first to demonstrate that it was possible to visualise different images using this method was the French painter G. A. Bois-Clair in 1692, but the use of this technique for stereoscopy was only applied in 1903 by Frederick E. Ives and later developed by his son Herbert E. Ives.A common application of this technique is holographic cards, also known as colorgrams. These cards feature several different images printed under a parallax barrier, so that an observer can visualise different images by tilting the card slightly. Schreck (2008) explains on his website how to create this type of card using image processing programmes and an inkjet printer.

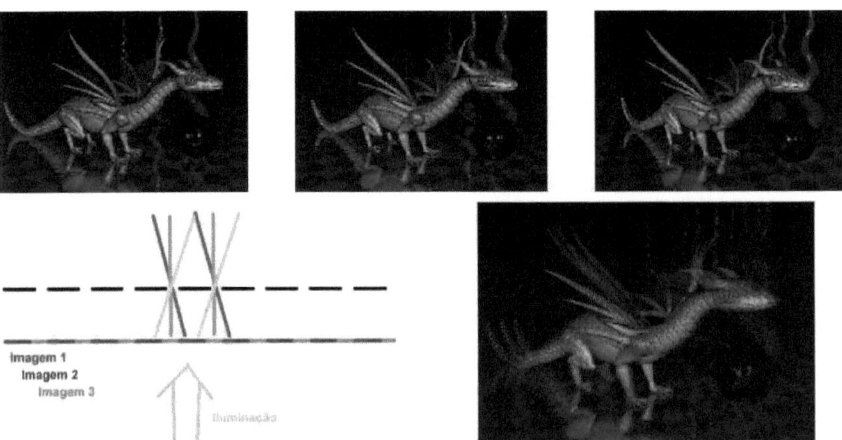

Figure 28 - Holographic card created by Schreck (2008). At the top are the initial images, on the left the schematic of how this technique works and on the right the final result.

Lenticular lenses

This technique uses the same principle as parallax barriers, directing the rays of light so that each eye receives the respective left and right image, but lenses are used instead of simple barriers (ROBERTS, 2003). The first step along these lines was taken by Gabriel M. Lippmann in 1908, with integral photography, in which a series of "fly's eye" lenses were placed on a special photograph, allowing it to be viewed from different points of view. A few years later, in the 1920s, several researchers including Herbert E. Ives continued to work on a simplification of Lippmann's method and proposed the use of a series of concatenated lenses, called a lenticular lens sheet. The great advantage of this way of grouping the lenses is that optically it becomes very similar to parallax barriers and therefore makes it possible to take advantage of much of the research used for the barriers.

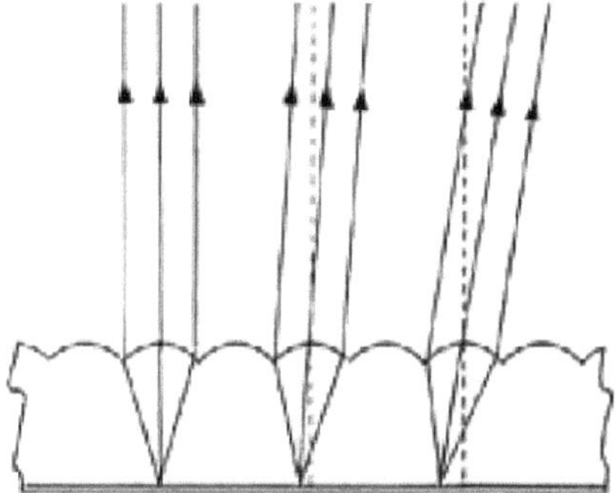

Figure 29 - Schematic of the optical behaviour of lenticular lenses (ROBERTS, 2003).

Volumetric Displays

A different approach to the autostereoscopy problem is the category of volumetric displays. In this type of solution, instead of trying to get a pair of planar images to visualise depth, the idea is to reconstruct a copy of the objects in space, so that an observer can visualise the object with depth, as this information will be naturally present. According to Halle (1997), in order to fulfil this task, solutions of this type need to somehow fill or sweep a volume in space. Several different technologies have emerged for this purpose, but they all have major limitations and this is still an area of intense research (WIKIPEDIA, 2008). One of the most recent representatives of this technology is a project developed by the Institute for Creative Technologies (ICT) at the University of Southern California. In this project entitled "Rendering for an Interactive *360°* Light Field Display", the researchers managed to create a volumetric display in which several simultaneous observers can visualise the reconstructed object.

But one of the great differences between this technology and other volumetric display solutions is the possibility of occluding the opposite part of the object, which would be hidden if it were a real object (JONES et al., 2007). The solution consists of using an overhead projector aimed at a mirror inclined at 45° which reflects the projection towards the observers. The mirror rotates at a frequency of approximately 20Hz and the projector is specially modified to be able to display around 4800 monochrome frames per second.

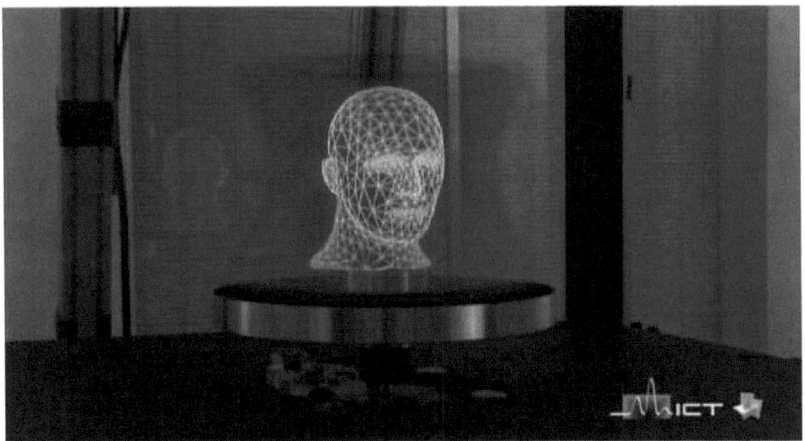

Figure 30 - Photograph of the solution developed by the ICT. Note the detail of the occluded back of the reconstructed head reflected in the glass (JONES et al., 2007).

2.2.13 Alternating Stereo Pair (Wobble/Wiggle Stereo)

This technique allows scenes to be visualised with a sense of depth, but it cannot be considered a stereoscopic solution because its operation is not based on the disparity of the images obtained by each eye, but rather on the principle of motion parallax.

An image in this format is actually an animation, which is made up of at least two images depicting the same scene from two slightly different points of view. By displaying these images in a cyclical and alternating manner, an observer gets the same feeling of parallax as when looking at the real world by moving their head slightly. This technique is already roughly described in a 1996 patent submitted by Benjamin R. Halpern (US PATENT 5524188). The patent deals with the technique in a more comprehensive way, describing various ways of exploiting parallax as a way of enabling the sensation of depth. The person responsible for producing the first digital versions on the internet and subsequently popularising this technique was Jim Gasperini (2002), who even provides some tips on his website for those wishing to create this type of animated image.

Figure 31 - Images used by Gasperini (2002) in an alternating stereo pair animation.

2.3 Stereoscopy market

Especially for applications that don't use HMDs, stereoscopy is still a topic that is looking for further consolidation on the market. Seasonally, a stereoscopic solution that enables popular use gains fame and several companies start manufacturing the equipment, but after a few years of excitement the technology recedes into the background and waits for the next big explosion in the stereoscopic market. Consequences of this market behaviour are the various products created at different times, lists of which can be found online at specialised sites such as Stereo3D.com and Stereoscopy.com.

In the late 1990s and early 2000s, the hot products were shutter glasses, which often came bundled with 3D graphics cards and promised that any game with three-dimensional rendering could be viewed with a sense of depth. Although at the time major hardware producers also joined in the excitement, what ended up happening was that the glasses never became more than a niche product and over time only a few manufacturers remained. One of the possible reasons for the decline of shutter glasses was the popularisation of LCD (Liquid Crystal Display) monitors, which, due to the way they update the screen, are not compatible with this type of stereoscopic solution (WEINAND, 2005a), or at least they weren't at the time.

Following the collapse of the shutter glasses market, one of the companies responsible for keeping this market alive is NVidia (2008), which has an active driver project for stereoscopic visualisation. With these drivers installed on a computer with a compatible operating system and graphics card, the user can select one of several forms of stereoscopic solutions that define the way the graphics card renders any application with virtual environments, regardless of whether the application was created with stereoscopic support in mind.

One line of equipment trying to reach the general public are autostereoscopic monitors and polarised monitors. The first type uses the same technique as lenticular lenses, dividing the monitor into vertical strips, where half of the strips display the left image and the other half the right image (WEINAND, 2005b). However, these monitors have some problems:

- Ordinary 2D images cannot be displayed because of the lenticular lenses;
- The images have half the horizontal resolution of an ordinary monitor;
- The viewer must be positioned in a narrow band in front of the monitor, otherwise the image will not be seen correctly.

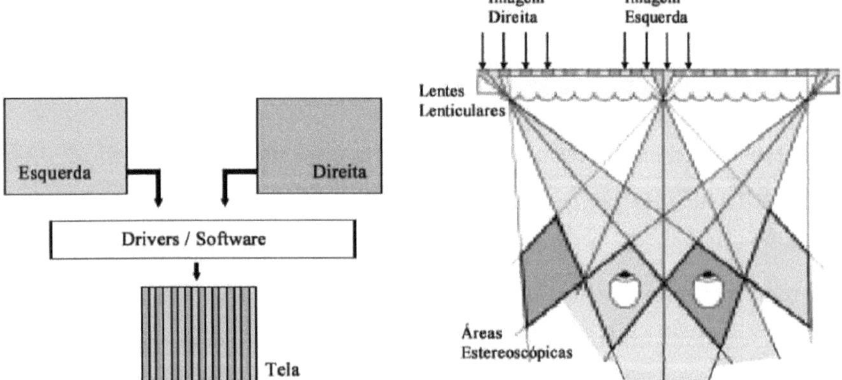

Figure 32 - Schematics of how autostereoscopic monitors work. Junction of the left and right images (left) and what happens when observing the monitor (based on WEINAND 2005b).

Polarised monitors use light polarisation technology, allowing an observer wearing glasses equipped with appropriate filters to view the image produced on the screen stereoscopically. One such model is the Trimon produced by Zalman (2008), which allows 2D and 3D images to be viewed in a wide horizontal field of vision, but is limited to approximately 10° of vertical movement of the observer. The main limitation of this type of solution is the need for the viewer to wear special glasses.

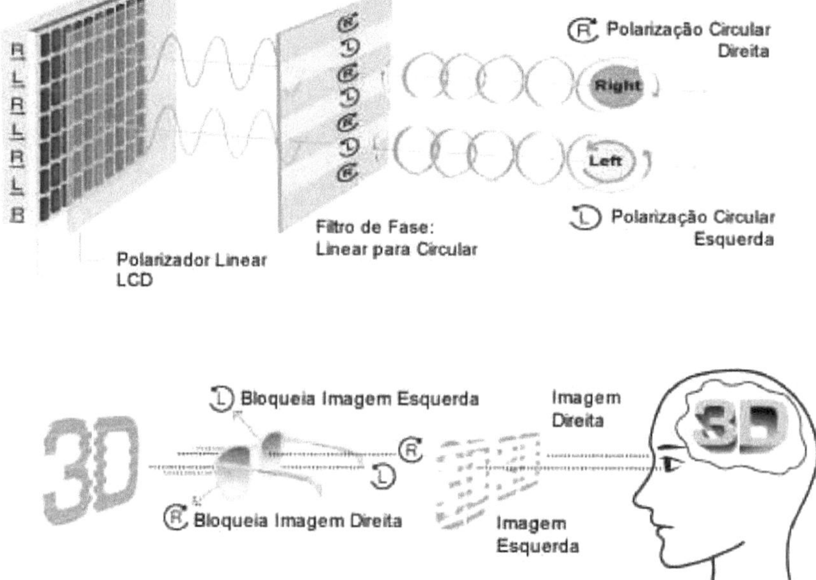

Figure 33 - Schematics of how polarised monitors work. Generation of polarised images (top) and how the observer views the images stereoscopically (based on ZALMAN, 2008).

In addition to domestic use, the cinema industry is increasingly exploiting stereoscopy, with more films and cinemas equipped with 3D theatres. The annual "Stereoscopic Displays and Applications" conference, which holds 3D film screenings, has been presenting increasingly large lists of films (WOODS, 2008a). Woods (2008b) shows that major production companies such as 20th Century Fox and Warner Brothers are taking risks in this niche, thanks also to the process developed by David Geshwind (2006) that makes it possible to convert a 2D film to be shown in 3D, although the process is still too slow for films to be released simultaneously in 2D and 3D.

Another Hollywood production driving this phase of stereoscopic growth is Avatar (2009). Written and directed by James Cameron, Avatar is the first blockbuster to exploit stereoscopy from its conception. As well as being produced with support for 3D cinemas in mind, the film's advertising was also aimed at encouraging viewers to watch the film in 3D. The film is the biggest box office success of all time, grossing over US$2.6 billion worldwide three months after its release date (IMDB, 2010). Although it is difficult to estimate the contribution of stereoscopy to Avatar's success, one irrefutable contribution is the ticket price. The difference between the ticket price for an ordinary cinema and a 3D cinema has inflated the film's gross revenue, as audiences are willing to pay more for a 3D cinema. In any case, this commercial success further emphasises stereoscopy, promoting the visual effects of this technology to an audience of different ages.

2.4 Virtual and Augmented Reality Works

Stereoscopy has been present in the field of virtual and augmented reality since HMDs began to support the display of different images for each eye, as the display of a stereo pair is a basic requirement of stereoscopy. Despite this, many of the projects involving stereoscopy are experimental. Below are some projects that involve both concepts in some way, starting with INTERLAB's AVTC.

2.4.1 Augmented Virtuality Tele-Conferencing (AVTC)

As described in other chapters of this work, AVTC is actually an architectural proposal for teleconferencing using the concept of augmented virtuality. Stereoscopy is applied as another element that increases the feeling of telepresence for users of this type of system. Figure 34 shows the AVTC architecture and then each module is detailed in a few words.

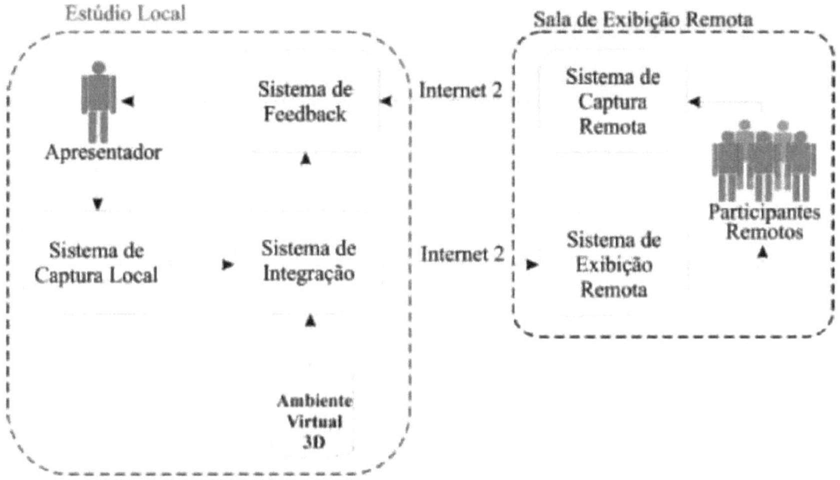

Figure 34 - AVTC conceptual architecture (SISCOUTTO, 2003).

In AVTC, users are divided into presenters and remote participants, who have different responsibilities within the system (SISCOUTTO, 2003). The presenter is the person whose image is captured by video and inserted in the form of a video-avatar within the virtual environment. Participants visualise the presenter within the virtual environment, with or without stereoscopy, and can interact via voice. In the studio where the presenter will be located, the following modules are present:

- **Local Capture System** - Responsible for capturing the presenter's actions via video, sound and movement commands;
- **Integration System** - Unites the data captured from the presenter, including the video avatar, with the virtual environment to form the scene that the participants visualise;
- **Feedback System** - Receives the data from the integration between the presenter and the virtual environment, adding this information to the voice and video captured from the participants. This information is passed on to the presenter.

The other point of the system is the exhibition room, where the participants are present and which contains the following modules:

- **Remote Capture System** - Obtains sound and video data from participants to be sent to the presenter;
- **Remote Display System** - Part responsible for displaying to participants the result of the integration between the virtual environment and the presenter's video avatar.

(a) (b) (c)

Figure 35 - Images of the AVTC architecture prototype. Left view (a), right view (c) and the stereo result for shutter glasses (b) (SISCOUTTO 2003).

Based on the knowledge acquired with this AVTC prototype, INTERLAB is continuing along the line of video-generated stereoscopic avatars with the Video-Avatar project, which is part of the TIDIA-AE distance learning project, funded by FAPESP.

2.4. 2CyberMED

The CyberMED project is made up of a set of libraries aimed at simplifying the development of virtual reality applications for the medical field. CyberMED has features such as: creation and manipulation of the virtual environment, collision detection, deformation of three-dimensional models, haptic interaction, evaluation of user training and stereoscopic visualisation (SANTOS; MACHADO, 2008).

CyberMED's libraries are based on OpenGL, an open standard of libraries that make it easier to manipulate graphics cards, speeding up the development of 3D applications.

In terms of visualisation, CyberMED has the usual planar output and three other types of stereoscopic output: anaglyph, colour anaglyph and shutter glasses.

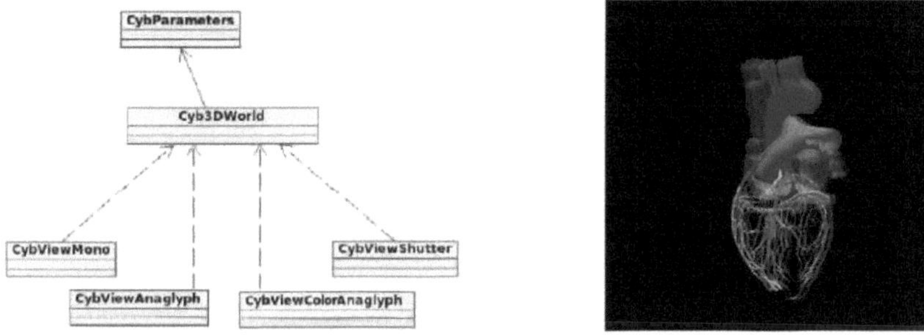

Figure 36 - CyberMED's visualisation class architecture and the result in a coloured anaglyph of an image of a heart (SANTOS; MACHADO, 2008).

2.4.3 IllusionHole

The IllusionHole is an application of stereoscopy using shutter glasses, but this solution has a special feature: it allows multiple users to visualise the same object in 3D from different points of view (KITAMURA et al., 2001). To achieve this effect, the IllusionHole consists of a flat surface like a table in which a circular hole is drilled. A few centimetres below this foreground is another surface on which the images that each user sees are projected, as shown in Figure 37.

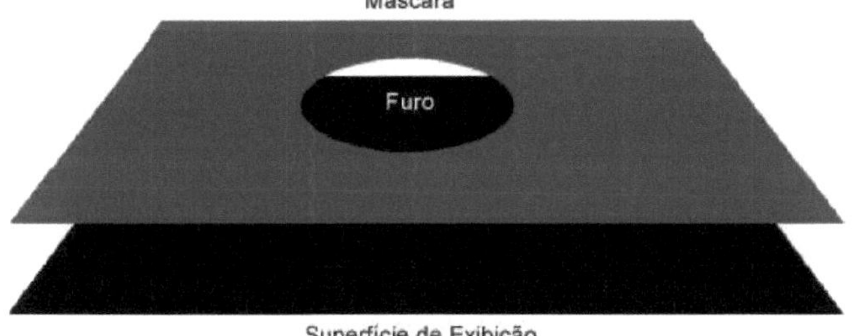

Figure 37 - IllusionHole assembly diagram (based on KITAMURA et al., 2001).

Additional equipment in this solution are trackers attached to the shutter glasses, used to identify the position of each observer in relation to the projection surface. With this information it is possible to generate images specific to the user's current position, so that users have the sensation of looking at the same object.

(a) Visão do ponto de vista de um quarto observador

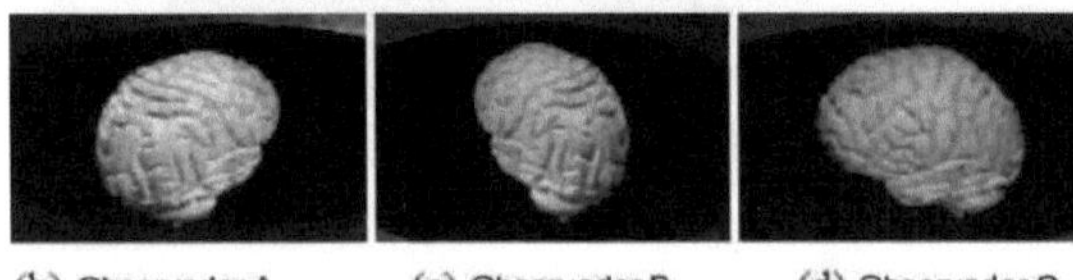

(b) Observador A (c) Observador B (d) Observador C

Figure 38 - IllusionHole in action and the observers' points of view (based on KITAMURA et al., 2001).

CHAPTER 3

METHODOLOGY

The initial step and foundation of the research is a literature review on the state of the art of stereoscopy. Due to the breadth covered by this area of knowledge, as described in item 2, this work encompasses a particular section of this topic, involving technology and equipment that enables the application of stereoscopy in computer graphics.

This literature search provides the theoretical framework for the next two stages, which are carried out in parallel: selecting the techniques to be evaluated and defining the tests to be applied.

Then, using prototypes based initially on the AVTC project (described in section 2.4.1), the test system is developed. With each new version of the prototype, characteristics such as performance and flexibility for incorporating stereoscopic techniques are evaluated. The final version of the system cannot restrict future solutions and tests, either because of platform restrictions or limitations of the language used for development.

With the test system implemented and functional, the next stage is to refine the tests, starting from an initial version developed from the theoretical conceptualisation and gradually evolving. The tests are considered mature when the stereoscopic solution used and the resulting depth perception are the main factors influencing the observer's results.

3.1 Selection of Techniques

First stage based on theoretical research into stereoscopy - the selection of which stereoscopic solutions have potential in the area of virtual and augmented reality. The importance of this stage is highlighted by the history of stereoscopy, which was born as a resource to enhance the results obtained in photography and cinema. Because of this historical factor, many of the solutions are not suitable for applications with the characteristics required by VR and AR systems, such as real-time rendering and control of the viewpoint of the scene by the observer.

Another factor taken into account when selecting solutions is availability, because due to the constant research and investment in this area, some are still prototypes or can only be obtained for a large sum of money. Although this additional factor restricts the scope of this research, it does not significantly affect its value, as the major experimental contribution of this work is the proposal, implementation and validation of the system used to assess the performance of observers with different stereoscopic solutions. Tests with other techniques and equipment can be carried out at a later date, without compromising the results obtained in this research.

The results of this selection process can be seen in section 4, which describes the experimental development.

3.2 Test Definition

The second front of action based on theoretical research - defining the tests to be carried out by observers using the system. The tests must make it possible to evaluate the stereoscopic solutions without other elements in the scene influencing the sensation of depth visualised by the observer. Section 2.1 details these other elements that influence depth perception.

The process for designing and developing each test was the same. Initially, qualities of a scene in VR or AR, viewed stereoscopically, that can be measured, whether quantitative or qualitative results, were determined. This stage resulted in the following list of characteristics:

- **Registration** - consistency between the position of elements in the virtual environment and the depth visualised stereoscopically;
- **Stability** - precision in the positioning of elements over time;
- **Field of View** - the freedom of the observer's angle of view in relation to the source of stereoscopic pairs, while still obtaining full stereoscopic perception;

- **Visual Sensation** - the quality of the depth perception obtained by the observer;
- **Visual fidelity** - correspondence between the stereoscopically observed image and the non-stereoscopically observed image;
- **Comfort and Ergonomics** - difficulty in adapting and discomfort felt by the observer when using the equipment required by the stereoscopic solution;
- **Response time** - the time required for the system to respond visually to a command from the observer;

- **Processing** - the cost, in terms of processing, to render a scene stereoscopically.

In order to facilitate an overview of the characteristics surveyed, the following Table 1 was created. In it, the characteristics are classified along two axes: the metric objectivity of the results and the specificity for the stereoscopy area.

Table 1 - Classification of the characteristics surveyed for the tests

	Specific	General
Quantity	Registration Stability Field of View	Response Time Processing
Qualitative	Visual Sensation Visual fidelity	Comfort and ergonomics

3.3 Test description

The tests were designed with a special focus on the Registration feature, which explores the relationship between position in the virtual environment and the observer's perception of depth. The tests have been grouped according to Table 1 to make it easier to identify the characteristic assessed by each test.

Specific - Quantitative

This category includes tests that seek to objectively measure the depth perception that the user observes when visualising a computer-generated environment stereoscopically. The aim of these tests is to identify the three-dimensional consistency that the observer has of the scene.

- **Depth (Registration)** - The observer must indicate the order of the elements present in the scene with respect to depth. Metric: Error in the order of the spheres;

- **Proportion (Registration)** - The viewer must adjust the scale and depth of one virtual piece so that it fits another. Metric: Scale and depth error;
- **Interaction (Registration)** - The observer must touch virtual elements distributed at different depths through another virtual object controlled by him. Metric: Consistency of the time taken to complete the test;
- **Relative position (Registration)** - The observer estimates the position of a virtual element in relation to the metric proposed by a virtual "ruler". Metric: Position error;
- **Position in space (Registration)** - The observer indicates the position in physical space of virtual points that they see "outside" the screen, due to stereoscopic positioning. This position is captured by cameras positioned perpendicular to and above the observer. Metric: Error between the relative positions indicated for the points.

Specific - Qualitative

This category includes tests that seek to subjectively measure the sensation of depth that the user observes when visualising a computer-generated stereoscopic environment. The aim of these tests is to identify the visual immersion that the observer has of the scene.

- **Consistency (Visual Sensation)** - An augmented virtual environment with a video avatar is generated and the observer must judge the degree of integration between the environment and the avatar. Metric: Score from 1 to 5;
- **Coherence (Visual Sensation)** - A virtual environment augmented with a video avatar is generated and the observer must judge the stability in the relative

position between the virtual and real elements of the scene. Metric: Score from 1 to 5;

- **Colour Fidelity (Visual Fidelity)** - Virtual objects of different colours are presented sequentially. The viewer must identify and report the colour of the object they are viewing. Metric: Colour classification errors.

General - Quantitative

This category includes tests that seek to objectively measure the software or hardware involved and are not directly related to stereoscopy.

- **Overhead (Processing)** - This test aims to objectively determine the increase in processing time added to the system due to the use of the stereoscopic technique. Metric: Increased processing time in per cent;
- **Performance (Processing)** - This test aims to objectively obtain the performance of the solutions on the hardware and software used for the tests. Metric: Number of frames per second.

General - Qualitative

This category includes tests that seek to objectively measure the software or hardware involved and are not directly related to stereoscopy.

- **Comfort and ergonomics** - After carrying out all the other tests, the observer rates the level of comfort when using the stereoscopic technique being judged. Metric: Score from 1 to 5.

3.4 Testing process

In order to determine a standard that each participant observer should follow in the testing process, a guide was created for carrying out the tests. This guide is presented in full in Appendix A. During the testing process, the researcher assists in navigating the system and the general progress of the tests, but does not make any gestures or comments that could influence the observer's choices.

3.4.1 Preparation

The order of the stereoscopic solutions and the tests carried out by each observer is selected randomly, avoiding the creation of trends due to familiarisation with the system and stereoscopy. The computer and equipment are set up in such a way that switching from one stereoscopic solution to another is as simple as possible, simply by changing the test system and asking the observer to change glasses.

The observer has an initial time to adapt to the stereoscopic solution and the testing system before the process begins. Once the observer is comfortable with the equipment and the system, the test session begins.

3.4.2 Execution

The tests are carried out in the order in which they were initially drawn, with all tests being carried out on the same stereoscopic solution before moving on to the next one. With each new test, the observer has an initial familiarisation attempt which is discarded, and only the results from the second attempt onwards are recorded. Once five valid results have been obtained in a test, the next test to be carried out by the observer is loaded.

3.4.3 Results

The results were recorded on an answer sheet (Appendix A) and in addition to the data obtained in the tests, information on the execution time and additional comments from the observers on the solutions were also included on this answer sheet. The unified presentation of the data can be seen below in section 5, as well as an analysis of the results found.

3.5 Analysing Test Results

Tests that do not require the participation of more than one observer, such as the processing tests, should be carried out by the researcher himself, while the other tests are carried out by different observers to enable greater generalisation of the results obtained and also to assess the system's performance in a real-life scenario. The researcher should record the data collected on the answer sheet, according to the process described above in item 3.4.

In quantitative tests, the data collected during the tests is used to calculate the performance of each observer with the different stereoscopic solutions. The main metric used to analyse this type of test is the error made by the observer, in the form of the squared deviation. The exception is the Interaction test, in which there is no error measure. For this test, the analysis is made in terms of the consistency of the observer's performance, i.e. the similarity of the results obtained by the observer when repeatedly performing a test with the same stereoscopic solution. The premise is that the player's performance does not vary much in this test, as long as they are viewing the scene with the correct depth perception.

Stereoscopic solutions can be compared by considering the observer's performance when using different stereoscopic techniques, including without stereoscopy. The efficiency of a solution can be given by comparing it with the observer's performance when doing the test without any stereoscopic equipment.

However, in qualitative tests in which observers rate each solution with a score, the comparison is more direct, with the observer's score being the determining factor in indicating which stereoscopic solution is better.

As far as possible, comparative tables of the solutions evaluated that unify the results obtained are created to facilitate data analysis. In addition to analysing and comparing purely experimental data, the solutions are also evaluated at a theoretical level, unifying and consolidating the conclusions and classifications of the stereoscopic techniques.

CHAPTER 4

EXPERIMENTS WITH STEREOSCOPIC TECHNIQUES

Considering the focus of this work on VR and AR systems, the stereoscopy solutions pre-selected to take part in the experimental stage must have the potential for this type of application. Therefore, the techniques need to fulfil some basic requirements for these systems:

1) Present images without too many distortions, which suggest to the viewer a cohesion between the real and virtual worlds;

2) Be suitable for use in an interactive system where the user controls the point of view.

3) The observer visualises the stereoscopic effect naturally, without using vision techniques.

Requirement number 3 aims to enable an observer with no prior knowledge of the solution to visualise the scene in three dimensions.

Table 2 below visually summarises which of these constraints each of the stereoscopic solutions meets. A dark square represents a constraint that the solution meets and a white square represents a constraint that is not met; squares of an intermediate colour mean a constraint that is met to a limited extent.

As shown in the table, considering the restrictions imposed, the solutions with potential for AR are: Anaglyph, 3D ColorCode, Light Polarisation, Shutter Glasses, Head Mounted Displays and Autostereoscopy.

In order to select the techniques that will actually be analysed, it is still necessary to assess the availability of equipment and technical information to implement the solutions. With these points in mind, the autostereoscopic solution still represents a high cost and the hardware is difficult to access when compared to other solutions. In addition, ColorCode 3D, due to patent restrictions and limited information available, turns out to be very difficult to implement correctly.

Table 2 - Restrictions of stereoscopic solutions

Solution	Restriction		
	1	2	3
Stereoscope	■		■
Stereo Slides	■		■
Stereo Pair	■	■	
Stereogram		■	
Pulfrich effect	■		■
ChromaDepth		■	■
Anaglyph	░	■	■
ColourCode 3D	░	■	■
Polarisation of Light	■	■	■
Shutter glasses	■	■	■
Head Mounted Display	■	■	■
Autostereoscopy	■	■	■
Alternating Stereo Pair*		░	■

* Conceptually it's not a stereoscopic solution, as both eyes see the same image.

4.1 Prototypes

Once the solutions to be implemented in the test system have been defined, the next step is the development of the software, which in addition to the modules with the stereoscopic solutions, also involves the implementation of a component capable of displaying images of environments and other virtual elements generated by computer graphics.

Some preliminary versions were implemented before reaching the definitive version of the system, which is used to test the stereoscopic solutions. All the prototypes follow the high fidelity methodology, i.e. they are as similar as possible to the result expected for the final system (HOM, 1998).

4.1.1 First version

The inspiration for the first version of the prototype was the AVTC project (described in section 2.3.1), which uses Java technology as the basis for implementation. In order to speed up the prototype development process, EnJine, a Java-based game engine (ENJINE, 2007), was used as a platform. The advantage of basing the implementation on an engine is the layer of abstraction it offers for handling common tasks such as: input and output devices, defining behaviour, collision detection, 3D rendering, among others (NAKAMURA et al., 2006).

EnJine is a project developed by INTERLAB that follows the ideology of opensource software, an additional positive point in an academic project that aims to be continued by other researchers. Although EnJine's main objective is to meet the needs of the introductory computer graphics course as a teaching tool, one of its applications is precisely as an experimental platform, thanks to its simplicity and robust architecture. From a technical point of view, EnJine only makes use of Java libraries, especially Java 3D, which is responsible for graphics generation. Because of this feature, it maintains a legacy of Java technology, which is its portability, i.e. the possibility of running the system on several different platforms without having to redo the coding. More information about EnJine, as well as its distributions and source code, can be found on the official website (ENJINE, 2007).

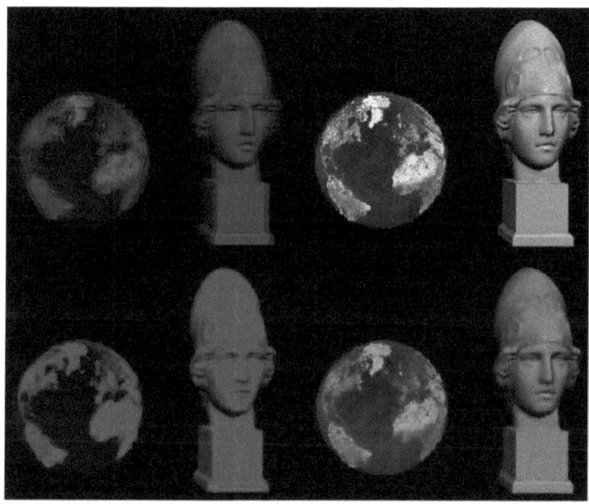

Figure 39 - Image generated by the Java prototype. Top right: normal rendering, bottom: red/cyan anaglyph pair, top left: complete anaglyph.

One problem that proved to be the biggest limiting factor of this version of the system was its lack of flexibility. Due to restrictions in Java itself and the libraries used, access to hardware devices is very slow or complex. For example, rendering in anaglyphs required four steps: rendering the scene in a temporary buffer in red, repeating the process in cyan, unifying the two temporary buffers and finally rendering the unified image on the screen. This task can be carried out more simply with the use of other technologies that offer better access to hardware, such as OpenGL (OPENGL, 2010).

Because of the interest in carrying out tests that exploit augmented reality, one of the system's needs is access to video cameras connected to the computer. Creating a video avatar as described in the AVTC project (SISCOUTTO, 2003) requires two video cameras to be accessed simultaneously, so that a coherent stereoscopic pair of the avatar is displayed to the viewer. Once again, Java's restrictions on hardware access made the task difficult. Using Java libraries such as the Java Media Framework (JMF, 2010) only one camera is accessed at a time, which is sufficient for home applications such as tele-conferencing, but makes it impossible to create a stereoscopic video avatar. The alternative used to get round this problem was to use the Java Native Interface (JNI, 2004) resource, which allows Java to access libraries in the machine's native language. In this case, the native library was OpenCV (OPENCV, 2010), which is a library written in C++ for computer vision. Although this alternative method allowed simultaneous access to two cameras, the cost in terms of processing to carry out this task proved to be very high and the cause of this additional cost was precisely the use of JNI.

A final limitation that was detected during the development of this prototype was due to Java 3D's own architecture. For applications that use ordinary three-dimensional rendering, without stereoscopy, Java

3D offers a resource that greatly facilitates the construction of the virtual environment, the scene graph (FOSTER, 2010). However, in this method of creating the virtual world, the developer's control of the rendering process is limited and, during the processing of stereoscopy techniques, fine control is often required, with low-level commands, so that the scene is rendered correctly.

4.1.2 Second version

After recognising the limitations mentioned in the previous section about the first version of the prototype, the next step was to establish a platform that could meet the flexibility and control required to implement stereoscopic techniques. The solution found was to migrate from Java technology to C++, which has the OpenGL and OpenCV libraries presented in the first version. With this decision, the portability of Java and the facilities offered by EnJine were exchanged for the low-level control and high performance offered by OpenGL and C++.

The coding of the system followed the Object Orientation paradigm, so some components were defined to make up the system as a whole. A class diagram in the UML (Unified Modeling Language) standard that represents the system as a whole is shown below:

To simplify the complexity of OpenGL-based programming, an additional library called GLUT (GLUT, 2010) was used. This library makes it easier to code tasks such as creating windows, the rendering loop and using the mouse and keyboard. One limitation that was not expected but became apparent during development in OpenGL was the lack of libraries for loading 3D models. Even though we searched extensively on the internet for libraries of this type, commonly called loaders, we couldn't find any that fully met the system's needs. The solution was to create our own loader based on GLM (GLM, 2010). This loader is the result of the efforts of several INTERLAB researchers.

The greater control of the rendering process offered by OpenGL has allowed stereoscopic techniques to be implemented without any major difficulties or limitations. This version supports virtual environment rendering for the following stereoscopic solutions: anaglyphs, shutter glasses, polarised glasses and HMD. As a curiosity, the alternate stereo pair (wobble stereo) has also been implemented.

Figure 40 below shows an image of the depth test rendered in anaglyphs, to illustrate the results obtained in this version.

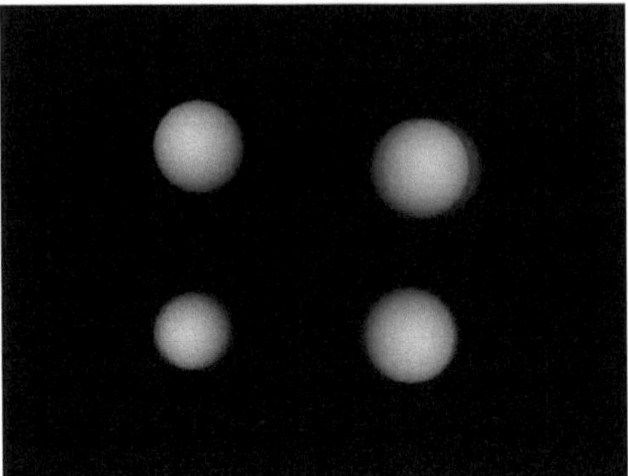

Figure 40 - Anaglyph rendered image of the depth test generated by the prototype in OpenGL.

4.2 Final version of the system

The final version is based on the second prototype and was developed using the same technology, C++ programming language with Object Orientation and the OpenGL library for 3D rendering. The main change from the prototype to the final version is its architecture. The organisation of the components that make up the system has been redefined to facilitate the expansion of existing modules and the creation of new ones. This simplified the work required to incorporate new stereoscopic solutions into the system, with different rendering modes, as well as additional tests. An overview of the architecture can be seen in Figure 41.

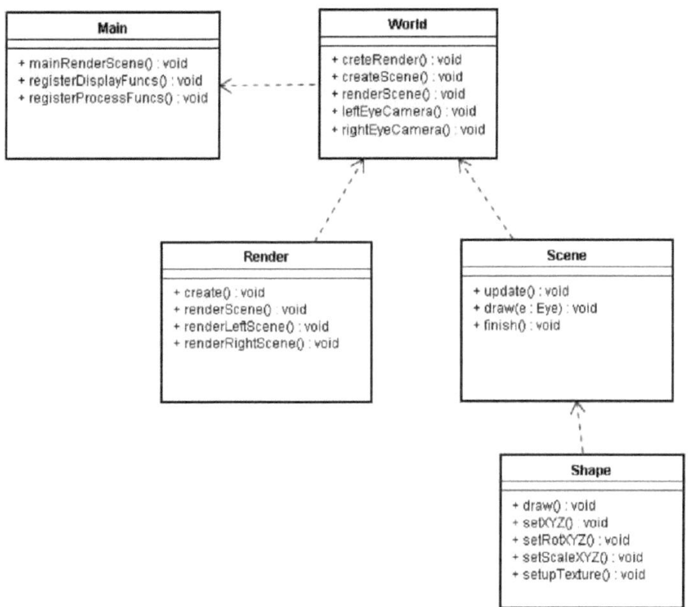

Figure 41 - UML diagram with main system components

Detailing each of the components:

Main

System root, defines the software initialisation functions. Due to the limitations of the GLUT component, several functions of this module are implemented outside the object-orientated standard, especially those that register image update calls and capture keyboard inputs.

World

Main module, responsible for managing the render modes and scenes for each of the tests. The keyboard inputs captured by **Main** are redirected to this component so that the user can select the **Render** and **Scene** types.

Render

This is an abstract component that must be implemented to create a new rendering mode. So far, five types of rendering have been implemented:

- **Normal** - without stereoscopy;
- **Anaglyphs** - rendering for red/cyan glasses;

- **HMD / Shutter Glasses** - implementation of the technique known as page-flip using the quad-buffer resource available in OpenGL;
- **Polarised glasses** - the creation of two windows that make up a stereo pair, for use with dual-output video cards and projectors;
- **Alternating** Stereo **Pair** - implementation of the wobble stereo technique to give a sense of depth.

Scene

It represents a 3D scene. Each of the system's tests is an implementation of this abstract component. The screen with the system's main menu is Scene 1 and the other tests are numbered sequentially, as illustrated in . The user of the system can select the scenes in the order they want by simply using shortcut keys.

Shape

This generic component represents an object in the virtual world. All the objects created in the scenes are specialisations of this component. As there is no concept of a scene graph in OpenGL, virtual objects need to be recreated in each rendered frame and this module simplifies the task of persisting objects from one frame to the next.

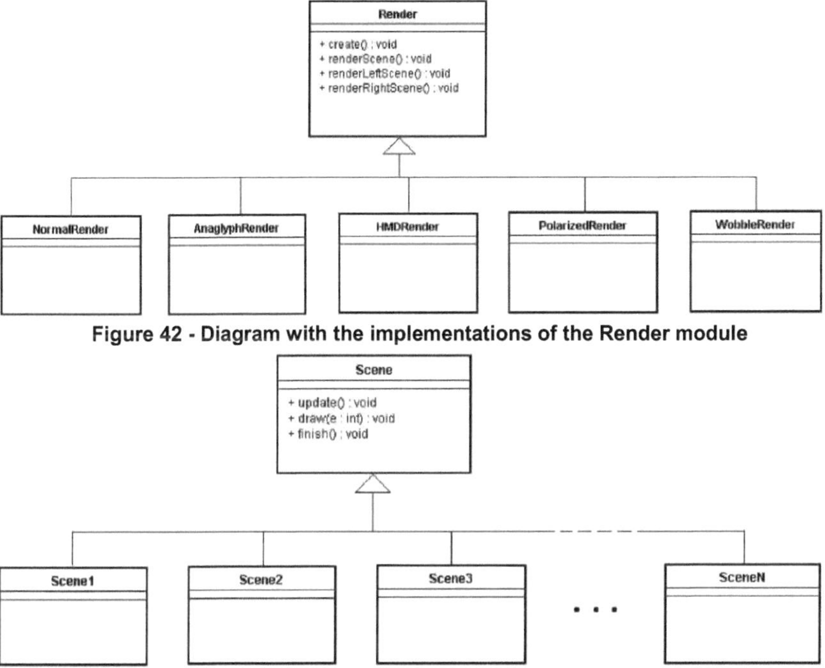

Figure 42 - Diagram with the implementations of the Render module

Figure 43 - Diagram illustrating the organisation of the scenes

1.3 Equipment

The equipment used to carry out the tests follows market standards and is easily accessible to the general public. Even the streoscopic solutions evaluated can be found without great difficulty in specialised shops.

Hardware

The computer running the system is a standard PC machine, and the following paragraphs show its configuration in more detail.

The video card is a Quadro FX1700, which offers the important quad-buffer feature, as well as the necessary performance. This feature is interesting because it allows the developer to implement stereoscopy with great control when using OpenGL.

The monitor is a Samsung 2233RZ, the model is important because the shutter glasses solution requires support for vertical refresh rates above 100Hz. This monitor was obtained together with the shutter glasses, model Nvision3D, both of which are sold as a package called the "3D Vision Bundle" marketed by the company Nvidia.

The anaglyph glasses were bought online and don't have any special features, they follow the red/cyan pattern.

Other components that are less relevant to the experience are the CPU, which has an Intel quad-core processor, 4gb of RAM and a standard SATA hard drive with a few hundred gigabytes.

Software

The operating system of the test computer is Windows XP 32bit and the only additional installation for running the test system are video drivers with support for quad-buffer stereoscopy in OpenGL. The Forceware driver version 192.0 and Nvision3D support (shutter glasses) for Windows XP, available on the Nvidia website, were used to run the tests.

1.4 Environmental conditions

The tests were carried out inside the laboratory, with the monitor positioned in a common bay. The laboratory environment in which the tests were carried out is structurally isolated from natural light, but no additional procedures were adopted to further control the conditions of the environment.

1.5 Test Implementation

The tests implemented for the execution of the experiments were based on the descriptions in item 3.3, however not all the tests were selected for the experimental test battery due to different factors, more details can be found in item 4.6.

If the reader wishes to evaluate the stereoscopic sensation experienced by the observers, all they have to do is obtain a standard red/cyan anaglyphic spectacle and compare the images of the tests without stereoscopic rendering and with stereoscopic rendering shown below.

1.5.1 Specific - Quantitative

Depth (Registration)

The system generates four spheres and positions each sphere in a quadrant of the screen, considering that there is a vertical and horizontal axis that cross in the centre of the screen. The spheres are positioned at different depths and have the same diameter. So that the observer cannot deduce which sphere is closest due to the perspective effect, it was disabled during this test.

Realisation: The observer must indicate the order of depth of the spheres, from closest to furthest.

Metric: Error in the order of the spheres.

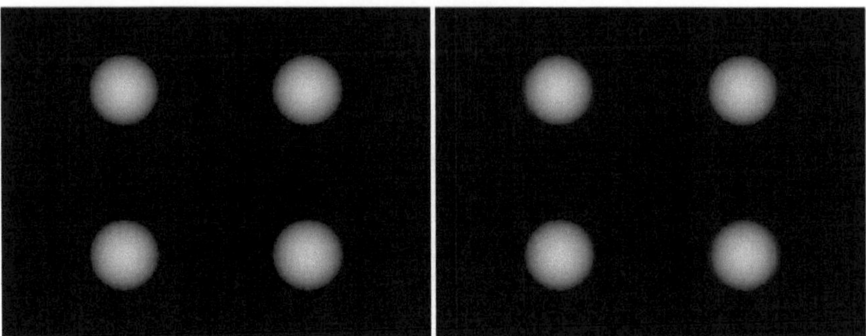

Figure 44 - Depth test. Image without stereoscopy (left) and in anaglyphs (right)

Proportion (Register)

Models of two pieces that fit together perfectly are loaded, each on one half of the screen, but each piece is assigned a different scale value. As the pieces are also positioned at different depths, the challenge for the viewer is to adjust the depth and scale of the piece on the right so that the pieces fit perfectly when positioned at the same depth.

Realisation: The viewer can increase or decrease the scale of the right part and also adjust its depth. When finished, the system indicates the scale error when positioned at the same depth as the part on the left.

Metric: Scale and depth error

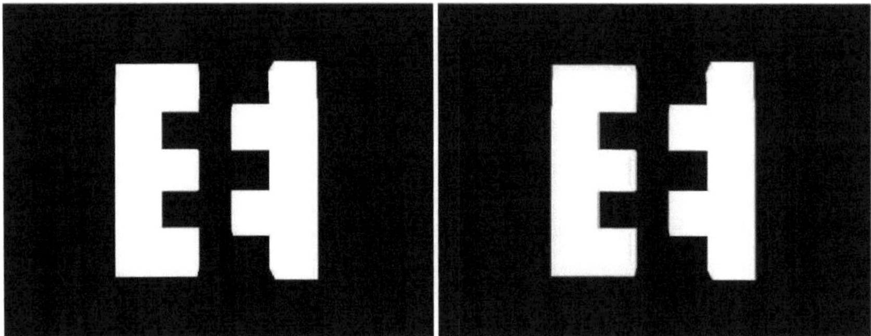

Figure 45 - Proportion Test. Image without stereoscopy (left) and in anaglyphs (right)

Interaction (Registration)

A ship is created that the viewer can move along different axes and rings are positioned in virtual space with equal sizes but different depths. The aim is to move the ship through space in order to pass through all the rings. So that the observer cannot deduce the depth of the rings due to the perspective effect, it was disabled during this test.

Realisation: The observer moves the ship in order to cross all the rings. When the last one is crossed, the system displays the time taken to complete the task.

Metric: Consistency of the time taken to cross all the hoops.

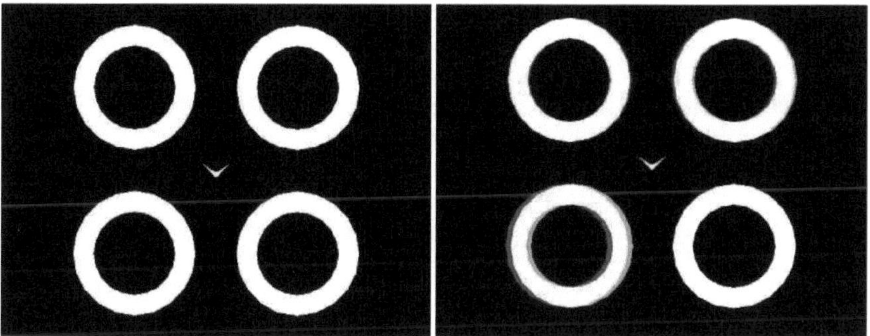

Figure 46 - Interaction Test. Image without stereoscopy (left) and in anaglyphs (right)

Relative Position (Register)

A sphere is generated near the centre of the screen, just above a "ruler" numbered from zero to fifty. The viewer must answer the position of the sphere in relation to the metric proposed by the "ruler".

Realisation: After evaluating the relative position between the objects, the observer answers the position of the sphere in the metric proposed by the ruler. At this point the test is finalised and the system reports the correct relative position.

Metric: Position error.

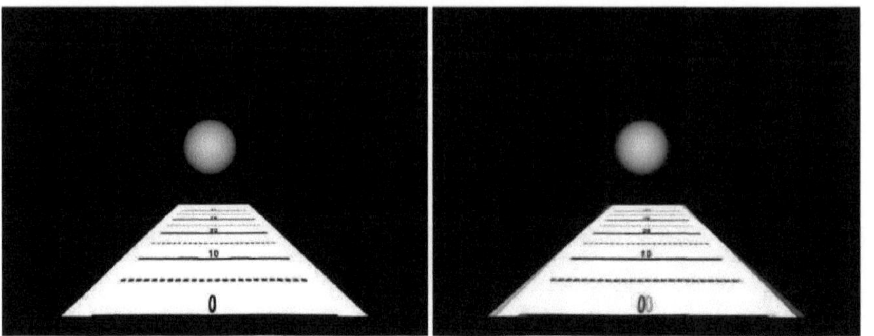

Figure 47 - Relative Position Test. Image without stereoscopy (left) and in anaglyphs (right)

Position in Space (Registration)

A geometric shape is generated in wireframe and positioned in the centre of the screen. The observer then points in physical space to the position where he sees the vertices of the geometric shape, considering that he will be seeing the object "outside" the screen due to stereoscopic positioning. Cameras positioned perpendicular to and above the observer will be capturing the relative position between each of the vertices identified by the observer.

Realisation: The test is started and the observer indicates the vertices in a predetermined order, always telling the system when they think they are pointing correctly at the vertex. The system then captures each vertex indicated by the observer and compares it with the vertices of the geometric shape generated.

Metric: Error between the relative positions of each vertex.

4.5.2 Specific - Qualitative

Consistency (Visual Sensation)

An augmented virtual environment with a video avatar is generated and the viewer must judge the degree of integration between the environment and the avatar.

Realisation: The observer must evaluate the result by giving it a score from 1 to 5.

Metric: Subjective score from 1 to 5.

Coherence (Visual Sensation)

An augmented virtual environment with a video avatar is generated and the observer must judge the stability in the relative position between the virtual and real elements of the scene.

Realisation: The observer must evaluate the result by giving it a score from 1 to 5.

Metric: Subjective score from 1 to 5.

Colour Fidelity (Visual Fidelity)

This test is only carried out to identify colour limitations resulting from the use of the anaglyph technique. Spheres of different colours are presented sequentially and the observer must identify and report which colour they are seeing.

Performance: The test is carried out in two sessions, first without stereoscopy and then using anaglyphs. With each sphere presented, the observer reports the colour seen.

Metric: The results of the two tests are compared and all classification errors are accounted for.

1.5.3 General - Quantitative

Overhead (Processing)

The aim of this test is to objectively determine how much more processing time is added to the system by using the stereoscopic technique.

Realisation: The processing time from one frame to another of an augmented virtual environment is calculated for each technique. These values are then described in relation to the processing time without any technique.

Metric: Percentage increase in processing time.

Performance (Processing)

The purpose of this test is to objectively determine the performance of the solutions on the hardware and software used for the tests.

Realisation: The number of frames rendered per second is calculated by the system and displayed to the researcher.

Metric: Number of frames per second.

1.5.4 General - Qualitative

Comfort and ergonomics

After carrying out all the other tests, the observer assesses the level of comfort when using the stereoscopic technique being judged.

Realisation: The observer must evaluate the stereoscopic solution.

Metric: Subjective evaluation.

1.6 Test selection

After a preliminary battery of tests to evaluate the system's operation, it became clear that the time needed to carry out the tests was too long. The average time for an observer to complete the tests was around two hours, while ITU-R, the reference for visual quality tests on televisions, recommends that the duration of a session should be thirty minutes (ITU-R BT.500-11, 2002).

The decision to reduce test execution time was to select a group of tests that have good cohesion and are sufficient to define a conclusive result. Subjective result tests were therefore eliminated from the battery, as they do not provide data that can be analysed directly.

The group of tests chosen to be carried out by observers are presented in section 4.5.1. The only exception is the Position in Space test, which cannot be included in the battery of tests because it is at too preliminary a stage to obtain consistent data.

CHAPTER 5

ANALYSING RESULTS AND CLASSIFYING TECHNIQUES

Initially, the system's processing, overhead and performance tests were carried out. The values were measured using a graphics processing evaluation module incorporated into the system itself. By activating this module, an annotation is displayed on the screen, informing you of the current rate of frames rendered per second and the average time elapsed between one frame and the next.

Secondly, the tests were carried out with the observers, following the process described in section 3.4. The observers performed the tests in three different ways: without the aid of a stereoscopic solution, with red/cyan anaglyph glasses and with shutter glasses. Although the system also supports stereoscopy with HMDs and polarised glasses, it was decided not to carry out the tests with these solutions, mainly due to the hardware they require. This differentiated hardware brings two complications for carrying out the tests. The first is the additional variables resulting from the change in the visual system, or in other words, the observer would have to view the test on a projected screen in the case of polarised glasses and view it through reduced LCD displays in the case of HMDs. The second is the time needed to carry out the battery of tests, which would increase significantly due to the change of equipment between one stereoscopic solution and another. This increase in time would further extend the duration of the test battery, which is already at the threshold of ITU recommendations, as described in section 4.6.

5.1 Results

The processing tests did not obtain significant results, probably due to the high capacity of the hardware available for the tests. Using the scenes created for the other tests, the difference in the results obtained between the stereoscopic solutions in the overhead and performance tests is confused with the fluctuations between each run.

A more suitable procedure for carrying out the processing tests is possibly to create a scene especially for this evaluation. The scene needs to be designed taking into account the capacity of the hardware that will be used during the tests, so that even if the difference in processing between the solutions is minimal, it is still possible to obtain relevant data for comparison.

Although the processing tests did not achieve interesting results for the research, the tests with observers generated a lot of material for comparing the solutions. However, better than presenting the raw data obtained in the tests is the data duly consolidated in the form of the error made by the observers. As described in section 3.5, the data is analysed based on the squared deviation of the observers' answers from the correct value presented by the system. In the case of the ship test, as there is no correct answer, the reference value used is the average of the observer's results, thus measuring the consistency of their performance. Table 3 below shows the deviation of the observers' responses for each of the tests individually.

With very few exceptions, the results with the greatest deviation were obtained when the observers were not using a stereoscopic solution, which indicates the directly proportional relationship between stereoscopic vision and performance in the tests.

One problem identified after collecting the data was the influence of parameters that are randomised by the system. In tests where the random factor has a strong influence on the scene, the variation in the deviation obtained in the absence of a stereoscopic solution is also very large. This can be seen in the results of the ruler and block tests, in which the random element influences the position, depth and size of the objects in the scene.

Table 3 - Quadratic deviation

		Sphere	**Ruler**	**Blocks**	**Ship**
Observer	Normal	140	2455	65,71	252,85
	Anaglyph	0	77	23,7	37,1
	Shutter	0	43	1,19	17,84
Observer2	Normal	148	509	38,43	411,34
	Anaglyph	0	306	14,03	20,18
	Shutter	0	111	3,7	100,86
Observer3	Normal	292	921	65,91	252,62
	Anaglyph	16	178	8,63	91,33
	Shutter	0	1127	4,86	0,66
Observer4	Normal	168	1228	19	485,71
	Anaglyph	52	715	34,51	443,28

	Shutter	0	286	13,66	233,11
Observer	Normal	100	1297	26,58	86,47
	Anaglyph	16	871	19,16	75,56
	Shutter	0	379	16,03	24,21
Observer6	Normal	84	128	150,22	217,41
	Anaglyph	4	371	16,55	26,91
	Shutter	0	39	86,9	63,15
Observer?	Normal	124	1348	35,07	158,44
	Anaglyph	136	1906	27,51	363,06
	Shutter	44	562	30,41	624,82

In order to carry out a comparative analysis of the stereoscopic solutions, it is interesting to organise this data relatively. Thus, the deviation values obtained in each test were normalised in relation to the deviation without the use of stereoscopy. Table 4 therefore shows the same data, but normalised in relation to the Normal reference value. Percentages above 100 indicate a higher deviation value than without stereoscopy and percentages below 100 a lower deviation.

Table 4 - Normalised squared deviation

		Sphere	**Ruler**	**Blocks**	**Ship**
Observer	Normal	100%	100%	100%	100%
	Anaglyph	0%	3%	36%	15%
	Shutter	0%	2%	2%	7%
Observer2	Normal	100%	100%	100%	100%
	Anaglyph	0%	60%	37%	5%
	Shutter	0%	22%	10%	25%
Observer3	Normal	100%	100%	100%	100%
	Anaglyph	5%	19%	13%	36%
	Shutter	0%	122%	7%	0%
Observer4	Normal	100%	100%	100%	100%
	Anaglyph	31%	58%	182%	91%
	Shutter	0%	23%	72%	48%
Observer	Normal	100%	100%	100%	100%
	Anaglyph	16%	67%	72%	87%
	Shutter	0%	29%	60%	28%
Observer6	Normal	100%	100%	100%	100%
	Anaglyph	5%	290%	11%	12%
	Shutter	0%	30%	58%	29%
Observer?	Normal	100%	100%	100%	100%
	Anaglyph	110%	141%	78%	229%
	Shutter	35%	42%	87%	394%

A comparison of the results obtained by each observer shows that the performance when using shutter glasses is consistently better than when using anaglyph glasses, although the difference is not striking. In particular, observer 4 had great difficulty using anaglyphs, with results close to or even greater than without the aid of any glasses. One possible hypothesis is that this observer has visual difficulties with the colour spectrum, on which the anaglyph solution is very dependent.

Another important aspect to note is that observer 7 is an exceptional case, as he has a clear difficulty in perceiving the depth of objects. Analysing his results, it is clear that none of the solutions stood out significantly compared to Normal, without stereoscopy. Only in the spheres and ruler test was the shutter glasses solution reasonably better, with a deviation just over fifty per cent lower than that obtained without any stereoscopic solution. A recurring comment after the tests was the visual fatigue and discomfort caused by using the anaglyph solution. In addition, some observers also commented on the improvement in stereoscopic perception after a period of adaptation, when using any stereoscopic solution, although this information was not clearly confirmed by the test results.

5.2 Classification

Initially, the aim of this section of the work was to classify stereoscopic solutions according to the data obtained during the tests, but this proved to be limited. Based on experimental data, the classification of solutions would be restricted mainly to test performance and there are several other factors that can be used for comparison and classification. For this reason, the proposal became a more comprehensive classification, also including theoretical information about stereoscopic vision and the technology involved in each solution.

The first point of discussion regarding the classification found in the literature is the separation between active stereo and passive stereo. This terminology is normally used to refer to stereoscopy by shutter glasses and by anaglyphic or polarised glasses, respectively. However, for this classification to accommodate other forms of stereoscopy, a more robust redefinition of what factors make a solution active or passive is first required.

In the context of this work, this classification of stereoscopic solutions refers only to the observer's part, in terms of the type of equipment they need to use to filter the stereo pair so that each eye only sees one of the images. The other part concerns the process of generating the stereoscopic images, more specifically, how the stereo pair is presented to the observer. The first part is called the **receiver** and the second the **sender** of stereo pairs. These concepts are better described in the following two sections (5.2.1 and 5.2.2), including examples of solutions representing the categories identified, both receivers and emitters. Table 5 below organises this classification visually, making it easier to understand how each solution fits in.

5.2.1 Stereo Pair Receiver

As a way of redefining the application of the terms active and passive stereo, this classification of stereoscopic solutions is proposed with the receiver of the stereo pair in mind. In this case, the term receiver refers to the equipment that somehow separates the stereo pair, sending one of the images to the observer's left eye and the other image to the right eye. This separation of the stereo pair between the two eyes is essential for obtaining depth through stereoscopy. The groups of stereoscopic receivers and how each of the solutions researched fits into this classification are presented below.

Dynamic (Active)

This category of receivers covers those whose state depends on time, i.e. the receiver's properties change depending on the instant at which we "photograph" its state. The main representative of this group are shutter glasses which, depending on the instant, block or allow visualisation through their lenses.

Static (Passive)

The complement of the Dynamic group. Static type receivers have a fixed state, which is independent of the instant at which it is checked. There are several examples of this group, in particular the historic anaglyph glasses and the more modern polarised glasses. Both have lenses with constant properties, which are always in the same state, free from the context of time. A special case of this group are HMDs, as the stereo pair transmitter and receiver coincide in the same device. HMDs have been categorised in the Static group because they are devices used by the observer to separate images through optical isolation, limiting the field of vision of each eye.

Absent

This group includes stereoscopic solutions that do not require a stereo pair receiver due to the characteristics of the transmitter. One example is autostereoscopic solutions, which, by the very nature of the techniques, do not require the observer to use equipment dedicated to separating the stereo pair.

5.2.2 Stereo Pair Transmitter

One way of classifying stereoscopic solutions is how they display the stereo pair to the viewer. In other words, the multiplexing mechanism that the emitter uses so that the images of the stereo pair do not mix, becoming a single indistinguishable image. Based on this classification criterion, three groups of stereo transmitters were observed: spatial, temporal and visible.

Space

This group includes transmitters that distinguish each image of the stereo pair through differences in position in space. A trivial example is autostereoscopic solutions, in which the left and right images are displayed from different positions and directions, to allow an observer to view a different image with each eye. Another member is HMDs, with independent displays for each image in the stereo pair.

Temporal

An alternative to the Spatial method is to display each image of the stereo pair at a different time. Thus, we have the Temporal category of transmitters, which display the left and right images at different times. The shutter glasses solution uses this approach, requiring monitors capable of displaying alternating images at high frequencies, so that it is imperceptible to the human eye. A special member of this category is the technique based on the pulfrich effect, which is based on the characteristic of human vision to perceive less intense light more slowly. Thus, the image visualised by one eye is always a few moments behind the other.

Visible

This category of emitters includes those that take advantage of some property of visible light to isolate the stereo pair. Two good examples are anaglyphs, which take advantage of the different wavelengths of the light spectrum, and polarisation, which uses the possibility of "aligning" the light as a way of isolating the images of the stereo pair.

Table 5 - Classification of stereoscopic solutions by type of emitter and receiver

	Dynamic (Active)	Static (Passive)	Absent
Space		HMD Stereoscope Stereo Slides	Autostereoscopy Parallel / Cross View Stereogram
Temporal	Shutter glasses	Pulfrich effect	Alternating Stereo Pair*
Visible	-	Anaglyph glasses Polarised glasses ChromaDepth	-

		ColourCode 3-D	

* Conceptually, it's not a stereoscopic solution because both eyes see the same image.

CHAPTER 6

AUGMENTED REALITY APPLICATION

Research into Augmented Reality is constantly looking for technologies and solutions that enable greater integration between real and virtual elements, possibly as a means of improving users' telepresence experience when using these systems. This search is spread across various segments involving AR, starting with the technology's capacity and precision and ending with the design of applications and interfaces that exploit more of the resources it offers.

An example of this evolutionary process is the technology used for the tracking and registration steps (AZUMA, 1997). Because the virtual elements of an AR system do not physically exist, their position in space needs to be determined from reference points of the real objects. One way of obtaining these reference points, which are necessary for tracking and registration, is to use fiducial markers. This type of solution is widely used in systems in which the user observes real scenes augmented by virtual objects via HMDs or mobile devices (BILLINGHURST et. al, 2001) (SCHMALSTIEG; WAGNER, 2007).

One of the first alternative markers is those detected by a library called ARToolKit (2010), developed at the end of the 90s. This library is able to determine the position and orientation of square markers that follow a black and white pattern. Several research groups have developed projects based on this library, one of which is the "Magic Book" (BILLINGHURST et. al, 2001), which demonstrates a way of augmenting a real book with virtual additions through AR. A few years later, another solution called ARTag (2010) was developed to overcome some of the problems and limitations of ARToolKit. Fiala (2004) carried out an extensive comparison of the two solutions, showing that ARTag has advantages in several characteristics such as false positives, marker confusion, loss due to occlusion, among others. Wagner and Schmalstieg (2009) present a review of this evolution, focusing on mobile devices, and show how various techniques have been developed in recent years to meet this AR challenge.

In the same way as technology, another segment that has demanded efforts in the area of AR is the applications of this technology, which are increasingly aimed at the general public. Researchers in the field of HCI (Human-Computer Interface) are looking for a way to create interfaces that get the most out of AR (GABBARD; SWAN, 2007), while others are working directly with users, such as Schmalstieg and Wagner (2007), who report on the development and experience of students using AR applications, aimed at education and entertainment, running on mobile devices.

The reason for raising the evolution of these fronts involving Augmented Reality is that both offer opportunities for the application of stereoscopy. In the first case, which involves tracking and recording elements, it is not stereoscopy, as discussed in this article, that results in the observer feeling a sense of depth, but the mathematical use of the disparity between the stereo pair. By comparing the image captured by two or more different cameras, with known position and orientation, it is possible to use the images captured by them to determine the position of the elements filmed in space. Pintaric and Kaufmann (2007) built a tracking system for VR and AR based precisely on this stereoscopic principle. The system detects and tracks infrared markers that are captured by video cameras in real time. The images obtained by the different cameras are processed mathematically following the concept of disparity, in order to determine the three-dimensional position of the elements in them.

In the second case, from an HCI point of view, stereoscopy provides an intuitive mechanism for increasing users' precision and speed when manipulating VR and AR systems. Because of the three-dimensional nature of these systems, the solutions used for two-dimensional interfaces introduce error and ambiguity arising from the additional degree of freedom. Teather and Stuerzlinger (2007) defined a good practice guide for creating 3D interfaces and also carried out some experiments to evaluate different types of interfaces. An interesting conclusion based on the experimental results is that stereoscopy has a significant impact on the accuracy of user actions.

5.3 Challenges of Stereoscopy in Augmented Reality

Stereoscopy and Augmented Reality are lines of research that go hand in hand, possibly due to the intrinsic relationship between them that an AR system ideally needs to be visualised stereoscopically in order to be faithful to the real world. In this sense, the challenges involving the two areas can be seen on two levels: one more conceptual, discovering innovative ways to unite these topics, and the other more practical, adding the other technology to projects and systems that involve only one of them.

An example of a project that adds stereoscopy to an AR system is presented by Peterson, Axholt and Ellis (2008), in which the authors use an AR system that increases the user's visual perception by showing virtual annotations associated with elements of a real external scene. The addition of stereoscopy facilitates the user's task of associating an annotation with a real element, inserting depth as another variable to help determine this relationship.

5.3.1 Avatar Video

One work in particular that aims to combine the two technologies is INTERLAB's Video-Avatar, which is part of FAPESP's TIDIA-AE project. The Video-Avatar module aims to enable an individual, in the case of the TIDIA-AE project a teacher, to teach remotely with a high degree of telepresence from the students' point of view. The teacher's video images captured in real time are incorporated into a virtual environment, forming a stereo pair that is then transmitted via the Internet, so that the student can observe the teacher and the environment with depth perception (TOKUNAGA, 2009).

CHAPTER 7

APPLICATION IN GAMES

The use of stereoscopic techniques is becoming an increasingly common strategy, although it has been exploited to a limited extent in the development of electronic games.

Although some studies indicate that this technology may cause discomfort for some players (HÃKKINEN et al., 2006), or may have no effect on actions carried out within an immersive environment (MCMAHAN et al., 2006), this feature has been present in some game engines (JACOBSON et al., 2005), in graphics devices for personal computers (NVIDIA, 2008) and in some console games, such as Sly3 (SLY3, 2009].

The first stereoscopic experiences in the area were limited to offering some support that allowed a small number of games to use a pair of 3D glasses (STEREO3D, 2010), which increased the player's sense of immersion. The next generation, on the other hand, brought an extensive list of titles, released in the 1980s and 1990s, which contained the feature natively (STEREO3D, 2010).

With the development of APIs such as Direct3D and OpenGL, which communicate directly with the graphics card driver as long as it has a 3D chipset, there has been a growth in the number of games that exploit stereoscopic vision in some way (STEREO3D, 2010).

The current generation of graphics devices has been designed to transform hundreds of games into a stereoscopic experience by modifying the operating system's video driver, even though these games were not originally designed to be used with this type of feature (NVIDIA, 2008).

As far as consoles are concerned, in the 1990s some games were released with 3D glasses for the Master System, from Sega, which produced the illusion of depth. The Playstation 2 adopted anaglyph technology, which was used in a non-mandatory way at certain times in some games (SLY3, 2009).

Still on the subject of consoles, in 2009, at the Consumer Electronics Show in Las Vegas, Sony showed prototype games for the Playstation 3 console (Wipeout HD, Grand Turismo 5 and Motorstorm) using stereoscopic effects.

5.4 Challenges of Stereoscopy in Games

Despite the diversity of styles and different levels of sophistication, it is possible to identify some common elements in electronic games that form part of their composition. As with these elements, it's interesting to first classify how stereoscopy fits into games.

7.1.1 Conceptualisation

Regardless of genre or level of graphical sophistication, computer games (or those run on any other platform) have characteristics in common, with elements that form part of their composition. Rollings and Morris (2004) categorised these elements into three groups. They are:

- **Integral** - vital for the game to function properly. The absence of such an element deprives the player of a substantial range of choices.
- **Chrome** - intensify the fun provided by the game, but do not affect gameplay. They emphasise the "atmosphere" that the game conveys to the player.
- **Substitute** - characterised by only providing an equivalent option for the player. They add little additional value to the game.

Both the first experiments applied to electronic games and the releases in the field shown recently at events feature stereoscopy as a visual accessory, which classifies it as an element of the chrome group, influential on aesthetics but not on gameplay.

An important step towards strengthening stereoscopy in the games market is the migration of this element from the chrome category to the integral category, i.e. vital for a full gaming experience. This aspect is discussed further in section 7.1.2 below.

7.1.2 Trends

The evolution of electronic games can be better understood by looking at, among other characteristics, the graphic quality and the various forms of interaction offered by today's games. The exponential increase in the processing power of home computers, combined with the cheapness of less conventional devices, has enabled manufacturers to explore the limits of this technology, adding a multitude of features to their games in order to make them more attractive to consumers.

In this sense, a large part of the efforts of researchers in the field are aimed at finding ways of increasing the level of immersion of players through visual quality. Resources such as stereoscopic vision, which makes it possible to visualise different levels of depth between the elements that make up the game's scenery, have proved to be an interesting alternative. However, regardless of the technique adopted, stereoscopy has been a means of simply enhancing the visualisation of the game's virtual scenery, used at specific times and on an optional basis.

This is a limited approach to the potential of stereoscopy. Considering that electronic games are interactive scenes augmented with rules that make it a game, any feature that adds degrees of freedom in the interaction between the player and the scene can be exploited directly in the gameplay, not just as a visual prop. The article by Tomoyose et al. (2009) presents this approach to using stereoscopy in games in more detail.

5.5 Implementing Stereoscopy in a *Game Engine*

The restrictions for implementing stereoscopy are strongly related to the hardware, software and stereoscopic solution chosen. As a means of organising the available options, the description is presented by stereoscopic solution.

Anaglyphs

One of the simplest techniques to implement, the platform chosen only influences the resulting performance, but no special features are required. This is an interesting option to implement in a prototype for testing, as anaglyph glasses are easy to obtain.

HMD and Shutters

The implementation of this technique is highly dependent on the hardware and software used, so the platform needs to explicitly support this type of solution. In the case of PC computers, there are basically two options: through the video hardware, which requires a video card with quad-buffer capabilities, or through the Nvidia drivers, which although they allow transparent stereoscopy for the developer, prevent them from having control of the stereoscopic rendering.

Polarised

Considering the usual method for implementing this solution, with two projectors each with a different polarising filter, the main requirement is hardware with two independent video outputs. The software must render the stereo pair and direct each image of the pair to a different video output. Just as a side note, the screen on which the images are projected also needs special attention, as it must be made of a material that maintains the polarisation of the light.

5.5.1 EnJine case

The implementation of a stereoscopic module in EnJine implies strong restrictions with regard to the software adopted, which follows Java technology. The Java3D library has the problem mentioned in section 4.1 of limiting the developer's control of the rendering process and even presents some problems when it comes to incorporating images captured by video cameras as real-time textures. Even with these considerations, a stereoscopic module for EnJine is feasible, always bearing in mind the performance limitations imposed by a higher-level, interpreted language such as Java.

An interesting way to create a stereoscopic module is to define an abstract class or interface that governs the pattern to be followed when implementing a technique. Following this format, each stereoscopic solution becomes a different component that can be invoked when necessary, allowing the rendering mode to be changed at runtime according to the desired solution.

Delving into more technical implementation details, the experience with the first prototype (item 4.1.1) resulted in a better understanding of how to develop the component corresponding to each stereoscopic solution. In the case of anaglyphs, for example, an interesting way of obtaining the stereo pair is to "cover" the two virtual cameras with transparent spheres that play the role of lenses. The sphere around the left camera only allows red light to pass through and the sphere positioned on the right camera allows cyan light to pass through. With this procedure we obtain the anaglyph pair without having to individually manipulate the images rendered in memory.

Developing a module for shutter glasses and HMDs only requires using the stereo feature in the Java3D documentation. Note that enabling this option is only possible if the video hardware supports the quad-buffer feature. Finally, a stereoscopic component for polarised glasses using projectors requires control of the computer's video outputs, so that the image obtained by the left camera is directed to one output and that obtained by the right camera to another output. The code to carry out this task can also be found in the Java3D documentation.

CHAPTER 8

CONCLUSIONS

As we were able to recognise from the theoretical research, stereoscopy is a very broad area of knowledge that involves various other fields in which the stereoscopic effect can be exploited, including art, medicine and computer graphics. In this work, it was decided to concentrate on stereoscopic solutions that allow images generated by computer graphics to be visualised with a sense of depth.

Initially, the literature research, which is the basis of this project, provided the resources for an important contribution: the creation of a consolidated and concise reference on the concepts, techniques and equipment used for stereoscopic visualisation, materialised in item 2 of this work.

Looking at it from the industry's point of view, it can be seen that, due to stereoscopy's history, starting with the arts, through photography and video, to computer science, stereoscopy has often been propagated as a niche product. The entertainment industry is attempting a new stereoscopic wave to establish this technology on the market. The number of 3D cinemas and games with stereoscopic support is growing every day, but an important factor is that they are becoming increasingly popular. This is the big difference between this new wave and the previous ones: the products are aimed at the general public, not just an accessory for "eccentrics". Major Hollywood productions such as the film Avatar and games from major producers such as Sony and its Playstation 3 are examples of the industry's efforts to popularise stereoscopy.

Although the industry is making a big effort to make a new visual leap through stereoscopy, research into the quality of stereoscopic techniques and the real sensation of observers is scarce, as is the study of other characteristics, such as the accuracy of the depth observed.

As a contribution to this gap and the main topic of this work, a system was proposed and developed to fulfil this need. This system allows different tests to be carried out with different stereoscopic solutions. Although only a few test modules and solutions were initially implemented, the system's architecture makes it easy to create additional modules. Although it was necessary to overcome some barriers, such as obtaining the appropriate hardware and migrating from the initial proposal to use Java and Java3D technology to C++ and OpenGL, the system was successfully implemented, showing good performance and the necessary flexibility for expansion.

A battery of tests was carried out with some expert observers, also as a way of validating the system and the methodology for creating the tests. The results obtained experimentally confirmed two essential aspects of the research: the possibility of creating tests that evaluate the stereoscopic effect and the fact that an observer has different results with different stereoscopic solutions.

On a more conceptual level, the work seeks to contribute to a more appropriate way of classifying stereoscopic solutions. Instead of limiting itself to the common and ambiguous terminology of active and passive stereo, the proposal is to separate it into two subsystems: the emitter and the receiver of stereo pairs. The emitter is the component responsible for displaying stereoscopic images and three categories have been defined: Spatial, Temporal and Visible. The receiver is the component used by the observer who is responsible for separating the left and right images for the respective eye. Three categories of receiver were proposed: Dynamic (Active), Static (Passive) and Absent.

Finally, as additional contributions, discussions are presented on the current state and trends of stereoscopy applied to AR and gaming. With regard to AR, the discussion seeks to demonstrate how stereoscopy can contribute to various challenges and objectives in this area, as well as simply being an additional resource available to achieve a greater sense of telepresence. In terms of games, with a focus on those involving computer graphics, the main thrust of the argument is the use of stereoscopy as an integral element of game mechanics, not just a visual prop.

5.6 Future work

This research opens up a number of opportunities for future work, both theoretical and practical. On a theoretical level, more effort is still needed to design tests that are less influenced by unpredictable factors, preferably forming a set that, regardless of the observers, when carried out without the aid of stereoscopy results in similar values. Still within the conceptual scope, the methodology of the test battery execution process needs to be consolidated in order to increase its robustness and reliability.

From a more practical point of view, there is the possibility of expanding the test system, either by incorporating new stereoscopic modules or implementing new tests. A more immediate task is to experiment with stereoscopic solutions that have already been implemented but were left out of the battery of tests, such as HMDs and polarised glasses.

Other fronts for future work are the application of stereoscopy in other areas, especially AR and games, which were part of the discussion in this research. In the field of AR, an interesting trend is the use of stereoscopy to increase the sensation of telepresence, particularly in video-avatars. One of the most interesting research fronts is the use of stereoscopy in games, which needs to cross the border of simple visuals to become part of the mechanics of games.

REFERENCES

3D EXPERIENCE. **3D Experience - A Thrilling Experience**. 2006. Available at: <http://www.freewebs.com/thrillzonein3d/3dpicturestosee.htm>. Accessed Nov. 2008.

ADAMS, G. A Bibliographical and Source Review of Stereoscopy. Anais do Museu Paulista, São Paulo, v.6/7, n.7, pp. 207-226, 2003.

AMAZON. **Linear Polarised 3D Glasses for IMAX movies and 3D Display**. Available at: <http://www.amazon.com/Linear-Polarized-3D-Glasses-Display/dp/B001 BG3A86>. Accessed on Nov.2008.

APO - American Paper Optics. **3D Glasses Online**. Available at: <http://www.3dglassesonline.com/>. Accessed in October 2008.

ARTOOLKIT. **ARToolKit Home Page**. Available at: <http://www.hitl.washington.edu/artoolkit/>. Accessed in Feb.2010.

ARTAG. **ARTag Home Page**. Available at: < http://www.artag.net/>. Accessed in Feb.2010.

AZUMA, R.T. **A Survey of Augmented Reality**. Presence: Teleoperators and Virtual Environments. vol. 6. pp 355-385, 1997.

BEENE, G. **Stereogram Overview**. Stereograms Information Centre. Available at: <http://www.garybeene.com/stereo/rds-over.htm>. Accessed in October 2008.

BELTRÃO, L. **An Approach to Stereoscopy via Green's Functions of Marriage Equations**. 2007. 84p. Dissertation (Master's Degree) - Fluminense Federal University, Niterói, 2007.

BILLINGHURST, M., KATO, H., POUPYREV, I. The MagicBook: A Transitional AR Interface. Elsevier Computers and Graphics, v.25, n.5, p. 745-753, November 2001.

BRITANNICA. **Louis Ducos du Hauron**. In: Encyclop^dia Britannica. Available at: <http://www.britannica.com/EBchecked/topic/172961/Louis- Ducos-du-Hauron>. Accessed in October 2008.

BURDER, D. **3D Images Ltd. 2005**. Available at: < http://www. stereoscopy. com/3d-i mages/photog raphy. html >. Accessed on Nov.2008.

BWH VENTURES. **Creating Custom Stereograms**. Available at: <http://www.eyetricks.com/stereograms/custom.htm>. Accessed November 2008.

CALIFE, D., TOMOYOSE, A., SPINOLA, D., BERNARDES JR, J., TORI, R. Robot ARena: infrastructure for applications involving spatial augmented reality and robots. In: SYMPOSIUM ON VIRTUAL AND AUGMENTED REALITY, 9., 2007, Petropolis. **Proceedings of the Symposium on Virtual and Augmented Reality**. Porto Alegre: SBC, 2007. p. 92-99.

COLLIER, I. **Stereo-Pair Images**. Iam Collier 3D Photographs 1993. Available at: <http://users.comlab.ox.ac.uk/ian.collier/3d/stereo.html>. Accessed in October 2008.

COLORCODE. **ColourCode 3-D**. Available at: <http://www.colorcode3d.com/whatis/index.htm>. Accessed Nov.2008.

COMBS, D. **A Walk Through the Roses of Reynolda Gardens in Pulfrich 3-D**. CombsMusic . Availableat : <http://www.combsmusic.com/RosesVideo.html>. Accessed in October 2008.

DAVY JONES LOCKER. **Dive Expeditions**. 2007. Available at: <http://www.techdivethailand.com/wreck-diving-photos.htm>. Accessed November 2008.

DEEB, S.; MOTULSKY, A. **Red-Green Colour Vision Defects**. GeneReviews. 2005.

Available at : <http://www.ncbi.nlm.nih.gov/bookshelf/br.fcgi?book=gene&part=rgcb>. Accessed in October 2008.

ENGDAHL, T. **3D Glasses and Other 3D Display Devices**. EPanorama.net. 1999. Available at : <http://www.epanorama.net/documents/pc/3dglass.html>. Accessed in October 2008.

ENJINE. **Java Game Engine**. enJine. 2007. Available at: <http://enjine.incubadora.fapesp.br>. Accessed in Feb.2010.

FEHN, C. 3D TV Broadcasting. In: SCHREER, O. KAUFF, P. SIKORA, T. **3D Videocommunication**. Chichester: John Wiley and Sons, 2005, p.23-38.

FIALA, M. **ARTag Revision 1: A Fiducial Marker System Using Digital Techniques.** National Research Council of Canada. 2004. Available at: < http://ecl. incheon. ac. kr/courses/ar7/articles/>. Accessed Feb.2010.

FOSTER, G. **Understanding and Implementing Scene Graphs**. Gamedev.net. Available at : <http://www.gamedev.net/reference/programming/features/scenegraph/>. Accessed in Feb.2010.

FRITSCH, D.; KADA, M. Visualisation Using Game Engines. In: ISPRS CONGRESS, 20, Istanbul, 2004. Istanbul: 2004. Commission 5, p. 621-625.

GABBARD, J.L., SWAN, J.E. Usability Engineering for Augmented Reality: Employing User-based Studies to Inform Design. IEEE Transactions on Visualisation and Computer Graphics, USA, v.14, n.3, p.513-525, 2007.

GASPERINI, J. **Time for Space Wiggle**. OCC Panoram. 2003. Available at: <http://3dculture.com/occ/Panoram_time4space_wiggle.htm>. Accessed in October 2008.

GESHWIND, D. **Inventor of 2D-to-3D Conversion Used for "Superman Returns" Unveils Improved 'Super-Fast' Technology for Next Time**. 2D- to-3D-Conversion: The Next Generation. 2006. Available at: <http://www.unipat.org/3dmg/PR2.htm>. Accessed in Feb.2010.

GLM. **GLM: an Alias Wavefront OBJ file library**. Frederic Devernay website, 2010. Available at: <http://devernay.free.fr/hacks/glm/>. Accessed in Feb.2010.

GLUT. **GLUT - The OpenGL Utility Toolkit**. OpenGL website, 2010. Available at: <http://www.opengl.org/resources/libraries/glut/>. Accessed in Feb.2010.

GRINBERG, V.; PODNAR, G.; SIEGEL, E.M. **Geometry of Binocular Imaging,** 1994. Availableat : <http://citeseer.ist.psu.edu/grinberg94geometry.html>. Accessed in October 2008.

HALLE, M. Autostereoscopic displays and computer graphics. **Computer Graphics, ACM SIGGRAPH,** v.31, n.2, pp. 58-62, May 1997. Available at: <http://web.media.mit.edu/~halazar/autostereo/autostereo.html>. Accessed Nov.2008.

HÃKKINEN, J., POLONEN, M., TAKATALO, J. AND NYMAN, G. Simulator sickness in virtual display gaming: a comparison of stereoscopic and non- stereoscopic situations. **Proceedings of the 8th conference on Human- computer interaction with mobile devices and services**. Helsing. New York: ACM Press, 227-230. 2006.

HOLLIMAN, N. S. 3D Display Systems. 2002. Available at: <http://citeseerx.ist.psu.edu/viewdoc/summary?doi=10.1.1.15.2146>. Accessed on Nov.2008.

HOM, J. **The Usability Methods Toolbox Handbook**. 1998. Available at: <http://www.idemployee.id.tue.nl/g.w.m.rauterberg/lecturenotes/UsabilityMethodsToolboxHandbook.pdf>. Accessed in Feb.2010.

IMDB. **Avatar**. Box Office Mojo. Available at: <http://boxofficemojo.com/movies/?id=avatar.htm>. Accessed in Feb.2010.

INITION. **Trivisio AR-vision Goggles/HMD**. Available at: <http://www.inition.co.uk/inition/product.php?URL =product hmd trivisio hm d seethrough&SubCatlD =16>. Accessed Nov.2008.

ITU-R BT.500-1. **Methodology for the subjective assessment of the quality of television pictures**. The ITU Radiocommunication Assembly. 2002. Available at : <http://www.dii.unisi.it/~menegaz/DoctoralSchool2004/papers/ITU-R BT.500-11.pdf>. Accessed on: Mar.2010.

JACOBSON, J., LE RENARD, M., LUGRIN, J.L. AND CAVAZZA, M. The CaveUT system: immersive entertainment based on a game engine. **Proceedings of the 2005 ACM SIGCHI International Conference on Advances in computer entertainment technology.** Valencia. New York: ACM Press, 184-187. 2005.

JAVA3D. **Java 3D Parent Project**. Java.net, 2010. Available at: <https://java3d.dev.java.net/>. Accessed in Feb.2010.

JMF. **Java SE Desktop Technologies - Java Media Framework API (JMF).** ORACLE Sun Developer Network, 2010. Available at:

<http://java.sun.com/javase/technologies/desktop/media/jmf/>. Accessed in Feb.2010.

JNI. **JDK 5.0 Java Native Interface-related APIs & Developer Guides**. Sun Microsystems, 2004. Available at : <http://java.sun.com/j2se/1.5.0/docs/guide/jni/>. Accessed in Feb.2010.

JONES, A.; MCDOWALL, I.; YAMADA, H.; BOLAS, M.; DEBEVEC, P. Rendering for an Interactive 360° Light Field Display. **Computer Graphics, ACM SIGGRAPH 2007 Papers Proceedings**, 2007. Available at: <http://gl.ict.usc.edu/Research/3DDisplay/>. Accessed Nov.2008.

KITAMURA, Y.; KONISHI, T.; YAMAMOTO, S.; KISHINO, F. Interactive Stereoscopic Display for Three or More Users. In: INTERNATIONAL CONFERENCE ON COMPUTER GRAPHICS AND INTERACTIVE TECHNIQUES, 2001, New York. **Proceedings of the 28th annual conference on Computer graphics and interactive techniques**. New York: ACM, 2001. pp. 231-240. Available at: <http://portal.acm.org/citation.cfm?id=383259.383285>. Accessed on: 23 March 2008.

KLEIN, A. **Stereoscopy.com - The Gallery**. Stereoscopy.com, 2008. Available at: <http://www.stereoscopy.com/gallery/index.html>. Accessed in October 2008.

KONDO, T.; SHIBASAKI, J.; ARITA-KIKUTANI, H.; MANABE, M.; INABA, R.; MIZUKI, A. **Mixed Reality Technology at a Natural History Museum**. 2007. Available at :

<http://www.archimuse.com/mw2007/papers/kondo/kondo.html>. Accessed on Nov.2008.

LPI. **Stereo Imagery: Anaglyphs**. Lunar Planetary Institute, Houston, Texas. 1999. Available at : <http://www.lpi.usra.edu/education/MarsMillennium/3dlesson.pdf>. Accessed in October 2008.

MCMAHAN, R.P., GORTON, D., GRESOCK, J., MCCONNELL, W. AND BOWMAN, D.A. Separating the effects of level of immersion and 3D interaction techniques. **Proceedings of the ACM symposium on Virtual reality software and technology**, Limassol New York: ACM Press, 108-111. 2006.

MILGRAM, P. et al. **Augmented reality: a class of displays on the reality- virtuality continuum**. Telemanipulator and Telepresence Technologies. 1994. Available at: <http://link.aip.org/link/?PSI/2351/282/1>. Accessed on: Oct.2008.

MINH, S.; FAZEKAS, K.; GSCHWINDT, A. **The Presentation of Three-**

Dimensional Objects with Single Image Stereogram. IEEE Transactions on Instrumentation and Measurement, v. 51, n.5, p.955-961, 2002.

NAKAMURA, R., BERNARDES JR, J., TORI, R. enJine: Architecture and Application of an Open-Source Didactic Game Engine. In: BRAZILIAN SYMPOSIUM ON COMPUTER GAMES AND DIGITAL ENTERTAINMENT, 5., 2006, Recife. **Digital Proceedings of the V Brazilian Symposium on Computer Games and Digital Entertainment**. Porto Alegre: SBC, 2006 p.1- 7.

NAVE, C.R. **HyperPhysics**. Georgia State University. 2006. Available at: <http://hyperphysics.phy-astr.gsu.edu/hbase/HFrame.html>. Accessed on: Oct.2008.

NVIDIA. **NVIDIA 3D Stereo**. NVIDIA website. Available at: <http://www.nvidia.com/object/3d stereo.html>. Accessed on: Oct.2008.

O'BRIEN, D. **Fog Day**. 2008. Available at: <http://gct13.com/e/20080224/fog-day-1 >. Accessed on: Oct.2008.

OPENCV. **Open Source Computer Vision**. OpenCV Wiki, 2010. Available at: <http://opencv.willowgarage.com/wiki/>. Accessed on: Feb.2010.

OPENGL. **OpenGL - The Industry Standard for High Performance Graphics**. OpenGL website, 2010. Available at: <http://www.opengl.org/>. Accessed on: Feb.2010.

PETTIGREW, J. D. The Evolution of Binocular Vision. In: Visual Neuroscience. Cambridge: Cambridge University Press, 1986, p. 208-222.

PETERSON, S.D., AXHOLT, M., ELLIS, S.R. Comparing Disparity Based Label Segregation in Augmented and Virtual Reality. In: VIRTUAL REALITY SOFTWARE AND TECHNOLOGY, 15., 2008, Bordeaux. **Proceedings of the ACM Symposium on Virtual Reality Software and Technology**. New York: ACM, 2008, p.285-286.

PHOENIX TROPICALS. **Growing Bermuda and Rye Grass in Phoenix Arizona**. Available at: <http://www.phoenixtropicals.com/grass.html>. Accessed on: Nov.2008.

PIEKARSKI, W. **Interactive 3D Modelling in Outdoor Augmented Reality Worlds**. 2004. Available at: <http://www.tinmith.net/wayne/thesis/piekarski-thesis.htm>. Accessed on Nov.2008.

PINTARIC, T., KAUFMANN, H. Affordable Infrared-Optical Pose-Tracking for Virtual and Augmented Reality. In: IEEE VR WORKSHOP ON TRENDS AND ISSUES IN TRACKING FOR VIRTUAL ENVIRONMENTS, 2007, Charlotte.

Proceedings of Trends and Issues in Tracking for Virtual Environments Workshop. Charlotte: IEEE, 2007, p.44-51.

RAINBOW SYMPHONY. **Proview Professional 3D Glasses**. 2000. Available at: <http://www.rainbowsymphony.com/pro3d.html>. Accessed on: Nov.2008.

RASKAR, R., WELCH, G. & FUCHS, H. Spatially augmented reality. In: INTERNATIONAL WORKSHOP ON AUGMENTED REALITY, 1., 1998, San Francisco. **Proceedings of the international workshop on Augmented Reality : placing artificial objects in real scenes**. Natick: A. K. Peters, 1998, p. 63-72.

REEL 3D. **3-D SlideViewers** . Available at: <http://www.stereoscopy.com/reel3d/slide-viewers-realist.html>. Accessed on: Nov.2008.

RMM. **An Illustrated Guide to Stereo Slide Mounting**. Rocky Mountain Memories2001 . Available at : <http://www.rmm3d.com/3d.encyclopedia/mounting/section3.html>. Accessed on: Nov.2008.

ROBERTS, D. **History of Lenticular and Related Autostereoscopic Methods**. Leap Technologies, Hillsboro, 2003. Available at: <http://www.microlens.com/pdfs/history of lenticular.pdf>. Accessed on: Nov.2008.

ROOLLINGS, A. AND MORRIS, D. **Game Architecture and Design: A New Edition**. New Riders, 43-45. 2004.

SANTOS, A.D.; MACHADO, L. S. Low-Cost Three-Dimensional Visualisation for the Development of Medical Applications. In: SYMPOSIUM ON VIRTUAL AND AUGMENTED REALITY, 10., 2008, João Pessoa. **Proceedings X Symposium on Virtual and Augmented Reality**, João Pessoa, 2008, p.179-182.

SCHMALSTIEG, D., WAGNER, D. Experiences with Handheld Augmented Reality. In: SYMPOSIUM ON MIXED AND AUGMENTED REALITY, 6., 2007, Washington. **Proceedings of the 2007 6th IEEE and ACM International Symposium on Mixed and Augmented Reality.** Washington: IEEE Computer Society, 2007. p.1-13.

SCHRECK, F. **Colograms - a simple way to create stereoscopic images**. Available at : <http://george.ph.utexas.edu/~schreck/schreck/cologram/>. Accessed on: Oct.2008.

SCREENDIGEST. **Digital Cinema building momentum as 3D drives market**... Available at : <http://www.screendigest.com/press/releases/pr 03 03 2010/view.html>.

Accessed in Mar.2010.

SILVA, R. **Optical See-ThroughHMD**. 2003. Available at: <http://virtual.lncc.br/~rodrigo/links/AR/node7.html>. Accessed on: Nov.2008.

SISCOUTTO, R.A. Proposal for a Three-Dimensional Teleconferencing Architecture Based on the Integration of Stereoscopic Avatar Video in a Three-Dimensional Virtual Environment. 2003. 192p. Thesis (Doctorate) - Polytechnic School of the University of São Paulo, São Paulo, 2003.

SISCOUTTO, R. A. SZENBERG, F. TORI, R. RAPOSO, A. B. CELES, W. GATTAS, M. Stereoscopy. In: TORI, R. KIRNER, C. SISCOUTTO, R. **Fundamentals and Technology of Virtual and Augmented Reality - Pre-Symposium Book, VIII Symposium on Virtual Reality**. Porto Alegre: SBC, 2006. c.13, p. 221-245.

SLY3, 2009. **SLY3: Honour Among Thieves**. Playstation. Available at: <http://www.us. playstation.com/ Sly3/>. Accessed on: May 2009.

STARKS, M. **Stereoscopic Imaging Technology**. Stereoscopic Imaging Technology, 3DTV Corporation. Available at: <http://www.3dmagic.com/articles/sit.html>. Accessed in October 2008.

STEREO3D.COM. **The independent 3D and Virtual Reality resource!** Available at: <http://www.stereo3d.com/3dhome.htm>. Accessed on: Mar.2010.

STEREOSCOPY.COM. **The World of 3D-Imaging**. Available at: <http://www.stereoscopy.com/>. Accessed on: Mar.2010.

STEUER, J. Defining Virtual Reality: Dimensions Determining Telepresence. The Journal of Communication, v.42, n.4, p.73-93, 1992.

TEATHER, R.J., STUERZLINGER, W. Guidelines for 3D Positioning Techniques. In: CONFERENCE ON FUTURE PLAY, Toronto, 2007. **Proceedings of the 2007 Conference on Future Play**. Toronto, Canada. 2007.

TOMOYOSE, A.N., SANCHES, S.R., TORI, R. Integrating Stereoscopy into Game Mechanics. **Proceedings: VIII Brazilian Symposium on Games and Digital Entertainment**. Rio de Janeiro, RJ, Brazil, 2009.

TOKUNAGA, D.M., SANCHES, R.R.S., TRIAS, P.L., BERNARDES J.A.L., NAKAMURA, R., TORI, R. Video-based microfacet-billboard avatar for educational immersive teleconference systems. In: SYMPOSIUM ON

VIRTUAL AND AUGMENTED REALITY, 11., Porto Alegre, 2009. **Proceedings of**

XI Symposium on Virtual and Augmented Reality. Porto Alegre, 2009.

TORI, R.; BERNARDES J.; NAKAMURA, R. Teaching Introductory Computer Graphics Using Java 3D, Games, and Customised Software: A Brazilian Experience. In: INTERNATIONAL CONFERENCE AND EXHIBITION ON COMPUTER GRAPHICS AND INTERACTIVE TECHNIQUES, 33, 2006, Boston. **Proceedings of XXXIII International Conference and Exhibition on Computer Graphics and Interactive Techniques**, Boston, 2006.

US PATENT 5524188. **Viewing three dimensional objects by rotational wobble about multiple axes**. Free Patents Online. Available at: <http://www.freepatentsonline.com/6687003.html>. Accessed on: Oct.2008.

US PATENT 6687003. **Method for Recording and Viewing Stereoscopic Images in Colour Using Multichrome Filters**. Free Patents Online. Available at: <http://www.freepatentsonline.com/6687003.html>. Accessed in October 2008.

VILELA, A.L.M. **Human Anatomy & Physiology**. Available at: <http://www.afh.bio.br/sentidos/sentidos1.asp>. Accessed on: Sep.2009.

WAACK, F. **Stereo Photography**. 2004. Available at: < http://www. stereoscopy.com/l i brary/waack-ch-5 .html>. Accessed on: nov.2008.

WAGNER, D., SCHMALSTIEG, D. History and Future of Tracking for Mobile Phone Augmented Reality. In: INTERNATIONAL SYMPOSIUM ON UBIQUITOUS VIRTUAL REALITY, 0., 2009. **Proceedings of International Symposium on Ubiquitous Virtual Reality** 2009, pp. 7-10.

WEINAND, L. **3D Shutter Glasses**. 3D Stereo Technology: Is it Ready for PrimeTime ? 2005. Available at : <http://www.tomshardware.com/reviews/3d-stereo-technology,1023-5.html>. Accessed on: Nov.2008.

WEINAND, L. **Autostereoscopic Displays**. 3D Stereo Technology: Is it Ready for Prime Time? 2005. Available at: <http://www.tomshardware.com/reviews/3d-stereo-technology,1023-5.html>.
Accessed on: Nov.2008.

WIKIPEDIA. **Volumetric Display**. Volumetric Display - Wikipedia the free encyclopedia. 2008. Available at : <http://en.wikipedia.org/wiki/Volumetric display>. Accessed on: Oct.2008.

WOLFGANG, C.; DANCY, M. **Polarisation Exploration**. WebPhysics. Available at:<http://webphysics.davidson.edu/physlet resources/dav optics/examples

/polarisation.html>. Accessed on: Oct.2008.

WOODS, A. **3D on Your PC**. Stereoscopic (3D) Imaging. 2005. Available at: <http://www.3d.curtin.edu.au/3D-PC/>. Accessed on: Oct.2008.

WOODS, A. **Stereoscopic Displays and Applications**. Stereoscopic Displays and Applications Conference Website. Available at: <http://www.stereoscopic.org/3dcinema/index.html>. Accessed on: Oct.2008.

WOODS, A. **The Illustrated 3D Movie List**. 3D Movie Website. Available at: <http://www.3dmovielist.com/list.html>. Accessed on: Oct.2008.

ZALMAN. **Trimon 2D/3D Convertible LCD Monitor**. Zalman, leading the world of Quiet Computing Solutions. Available at: <http://www.zalman.co.kr/ENG/product/Product_Read.asp?idx=219>. Accessed on: Nov.2008.

ZYDA, M. **From Visual Simulation to Virtual Reality to Games**. Available at: <http://csdl2.computer.org/persagen/DLAbsToc.jsp?resourcePath=/dl/mags/co/&toc=comp/mags/co/2005/09/r9toc.xml&DOI=10.1109/MC.2005.297>. Accessed on: 21 August 2008.

ZYGMUNT, P. YUNFENG, L. GREGORY, F. A New Look at Binocular Stereopsis. Vision Research, v.45, n.17, p.2244-2255, 2005.

ANNEX A

		1	**2**	**3**	**4**	**5**
Spheres	result					
	solution			1 - 2 - 3 - 4		
Ruler	result					
	solution					
Blocks	result					
	solution			10,0		
Ship	result					

		1	**2**	**3**	**4**	**5**
Spheres	result					
	solution			1 - 2 - 3 - 4		

Ruler	result					
	solution					
Blocks	result					
	solution	10,0				
Ship	result					

		1	2	3	4	5
Spheres	result					
	solution	1 - 2 - 3 - 4				
Ruler	result					
	solution					
Blocks	result					
	solution	10,0				
Ship	result					

Printed by Books on Demand GmbH, Norderstedt / Germany